Fundamentals of Nuclear Reactor Physics

Fundamentals of Nuclear Reactor Physics

E. E. Lewis

Professor of Mechanical Engineering
McCormick School of Engineering and Applied Science
Northwestern University

AMSTERDAM • BOSTON • HEIDELBERG • LONDON
NEW YORK • OXFORD • PARIS • SAN DIEGO
SAN FRANCISCO • SINGAPORE • SYDNEY • TOKYO

Academic Press is an imprint of Elsevier

Academic Press is an imprint of Elsevier
30 Corporate Drive, Suite 400, Burlington, MA 01803, USA
525 B Street, Suite 1900, San Diego, California 92101-4495, USA
84 Theobald's Road, London WC1X 8RR, UK

Cover Design: Joanne Blank
Cover Image © Mike Bentley/iStockphoto

This book is printed on acid-free paper. ∞

Library of Congress Cataloging-in-Publication Data
Application submitted.

British Library Cataloguing-in-Publication Data
A catalogue record for this book is available from the British Library.

ISBN: 978-0-12-370631-7

For information on all Academic Press publications
visit our Web site at www.books.elsevier.com

Working together to grow
libraries in developing countries

www.elsevier.com | www.bookaid.org | www.sabre.org

ELSEVIER   BOOK AID
International   Sabre Foundation

To Ann

Contents

Preface

This text is intended as a first course in the physics of nuclear reactors. It is designed to be appropriate as an introduction to reactor theory within an undergraduate nuclear engineering curriculum, as well as for a stand-alone course that can be taken by undergraduates in mechanical, electrical, or other fields of engineering who have not had a previous background in nuclear energy. Likewise, it is planned to be useful to practicing engineers from a variety of disciplines whose professional responsibilities call for familiarity with the physics of nuclear reactors.

Why a new book on reactor physics when a number of legacy texts are still in print? The better of these are well written, and since the fundamentals of the subject were already worked out at the time of their publication, they remain useful today. My conviction, however, is that for today's undergraduates and practicing engineers an introduction to reactor physics is better presented through both reorganizing and refocusing the material of earlier texts, and in doing that emphasizing the characteristics of modern power reactors.

Earlier textbooks most commonly have begun with the relevant nuclear physics and neutron interactions, and then presented a detailed treatment of neutron slowing down and diffusion in homogeneous mixtures of materials. Only in the latter parts of such texts does the analysis of the all-important time-dependent behavior of fissionable systems appear, and the dependence of criticality on lattice structures of reactor cores typically is late in receiving attention. To some extent such a progression is necessary for the logical development of the subject. However, both in teaching undergraduates and in offering continuing education instruction for practicing engineers, I have found it advantageous to present a quantitative but more general overview early in the text, while deferring where possible more detailed analysis, and also the more advanced mathematics that accompanies it, to later. Thus I have moved the treatment of reactor kinetics forward, attempting to inculcate an understanding of the time-dependent behavior of chain reactions, before undertaking the detailed treatment of spatial power distributions, reflector saving, and other topics dependent on the solution of the neutron diffusion equation. Likewise, the compositions of power reactor cores are incorporated into the discussion early on, emphasizing the interdisciplinary nature of neutronic and thermal design.

My intent in this text is to emphasize physical phenomena, rather than the techniques necessary to obtain highly accurate results through advanced numerical simulation. At the time the legacy texts appeared, computers were emerging as powerful tools for reactor analysis. As a result, the pedagogy in teaching reactor theory often emphasized the programming of finite differencing techniques, matrix solution algorithms, and other numerical methods in parallel with the analysis of the physical behavior of reactors, and thus extended the range of solutions beyond what could be obtained with paper and pencil. Now, however, high level programming languages, such as Mathcad® or MATLAB®, allow students to solve transcendental or linear systems of equations, to integrate differential equations, and to perform other operations needed to solve the preponderance of problems encountered in an introductory course without programming the algorithms themselves. Concomitantly, the numerical simulation of reactor behavior has become a highly sophisticated enterprise; one to which I have devoted much of my career. But I believe that the numerical techniques are better left to more advanced courses, after a basic understanding of the physical behavior of reactors has been gained. Otherwise the attempt to incorporate the numerical approaches employed in reactor design dilutes the emphasis on the physical phenomena that must first be understood.

By reorganizing and refocusing the materials along these lines, I hope to have broadened the audience for whom the text may be useful, in particular those advancing their professional development, even while they may not be taking a formal reactor physics course in a university setting. My goal is that the book may be read and physical insight gained, even though for lack of time or background some of the more mathematical sections are read and the results accepted without following each step of the development in detail. With this in mind, many of the results are presented in graphical as well as analytical form, and where possible I have included representative parameters for the major classes of power reactors in order to provide the student with some feel for the numbers.

Examples are integrated into the text's narrative, and a selection of problems is provided at the end of each chapter. The problems both enforce the concepts that have been covered and in some cases expand the scope of the material. The majority of the problems can be solved analytically, or with the use of a pocket calculator. In some cases, where multiple solutions or graphical results are called for, use of the formula menu of a spreadsheet program, such as Excel™, removes any drudgery that might otherwise be entailed. Selected problems require the use of one of the earlier mentioned high level computing languages for the solution of transcendental or differential equations. These are marked with an asterisk.

The preparation of this text would have been immensely more difficult if not impossible without the help and encouragement of many friends, colleagues, and students. Advice and assistance from the staff of the Nuclear Engineering Division of Argonne National Laboratory have been invaluable in the text's preparation. Won Sik Yang, in particular, has provided advice, reactor parameters, graphical illustrations, and more as well—taking the time to proofread the draft manuscript in its entirety. Roger N. Blomquist, Taek K. Kim, Chang-ho Lee, Giuseppe Palmiotti, Micheal A. Smith, Temitope Taiwo, and several others have also pitched in. Bruce M. Bingman and his colleagues at the Naval Reactors Program have also provided much appreciated help. Finally the feedback of Northwestern University students has been most helpful in evolving a set of class notes into this text. Most of all, Ann, my wife, has endured yet another book with grace and encouragement, while covering for me in carrying much more than her share of our family's responsibilities.

CHAPTER 1

Nuclear Reactions

1.1 Introduction

Albert Einstein's $E = mc^2$ relating energy to mass and the speed of light arguably is the most celebrated formula in the modern world. And the subject of this text, nuclear power reactors, constitutes the most widespread economic ramification of this formula. The nuclear fission reactions that underlie power reactors—that is, reactors built to produce electric power, propulsion for ships, or other forms of energy use—convert measurable amounts of mass to energy. Thus an appropriate place to begin a study of the physics of nuclear power is with the underlying nuclear reactions. To understand the large amounts of energy produced by those reactions in relation to the mass of fuel consumed it is instructive to introduce our study by comparing the production of nuclear power with that created by fossil fuels: coal, oil, or natural gas. Contrasting these energy sources, which result from chemical reactions, to nuclear energy assists in understanding the very different ratios of energy created to the masses of fuel consumed and the profound differences in the quantities of by-products produced.

Coal is the fossil fuel that has been most widely used for the production of electricity. Its combustion results predominantly from the chemical reaction: $C + O_2 \rightarrow CO_2$. In contrast, energy production from nuclear power reactors is based primarily on the nuclear reaction neutron + uranium-235 → fission. Energy releases from both chemical and nuclear reactions is measured in electron volts or eV, and it is here that the great difference between chemical and nuclear reactions becomes obvious. For each carbon atom combusted about 4.0 eV results, whereas for each uranium atom fissioned approximately 200 million eV, or 200 MeV is produced. Thus roughly 50 million times as much energy is released from the nuclear fission of a uranium nucleus as from the chemical combustion of a carbon atom.

For comparison, consider two large electrical generation plants, each producing 1000 megawatts of electricity (i.e., 1000 MW(e)), one burning coal and the other fissioning uranium. Taking thermal efficiency and other factors into account, the coal plant would consume approximately 10,000 tons of fuel per day. The uranium consumed by the nuclear plant producing the same amount of electrical power, however, would amount to approximately 20 tons per year. These large mass differences in fuel requirements account for differences in supply patterns. The coal plant requires a train of 100 or more large coal cars arriving each day to keep it operating. The nuclear power plant does not require a continual supply of fuel. Instead, after its initial loading, it is shut down for refueling once every 12 to 24 months and then only one-fifth to one-fourth of its fuel is replaced. Similar comparisons can be made between fossil and nuclear power plants used for naval propulsion. The cruises of oil-powered ships must be carefully planed between ports where they can be refueled, or tanker ships must accompany them. In contrast, ships of the nuclear navy increasingly are designed such that one fuel loading will last the vessel's planned life.

The contrast in waste products from nuclear and chemical reactions is equally as dramatic. The radioactive waste from nuclear plants is much more toxic than most by-products of coal production, but that toxicity must be weighed against the much smaller quantities of waste produced. If reprocessing is used to separate the unused uranium from the spent nuclear fuel, then the amount of highly radioactive waste remaining from the 1000-MW(e) nuclear plant amounts to substantially less than 10 tons per year. In contrast, 5% or more of the coal burned becomes ash that must be removed and stored in a landfill or elsewhere at the rate of more than five 100-ton-capacity railroad cars per day. Likewise it may be necessary to prevent nearly 100 tons of sulfur dioxide and lesser amounts of mercury, lead, and other impurities from being released to the environment. But the largest environmental impact from burning fossil fuels may well be the global warming caused by the thousands of tons of CO_2 released to the atmosphere each day by a 1000-MW(e) coal-fired power plant.

1.2 Nuclear Reaction Fundamentals

While an in-depth understanding of the physics of the nucleus can be a prodigious undertaking, a relatively simple model of the nucleus will suffice for our study of nuclear power reactors. The standard model of an atom consists of a very dense positively charged nucleus, surrounded by negatively charged orbiting electrons. Compared to the size of atoms, with diameters of roughly 10^{-8} cm, the size of the nucleus is very small, of the order of 10^{-12} cm. For modeling purposes we consider a nucleus to be made up of N neutrons and Z protons. Both are referred to as nucleons, thus the nucleus has $N+Z$ nucleons.

The number of protons, Z, is the atomic number; it determines an atom's chemical properties, while $N + Z$ is its atomic weight. Nuclei with the same atomic number but different atomic weights, due to different numbers of neutrons, are isotopes of the same chemical element. We refer to a nucleus as ${}^{N+Z}_{Z}X$, where X is the symbol used in the periodic table to designate the chemical element.

Reaction Equations

Nuclear reactions are written as

$$A + B \rightarrow C + D. \tag{1.1}$$

An example of a nuclear reaction is

$$ {}^{4}_{2}\text{He} + {}^{6}_{3}\text{Li} \rightarrow {}^{9}_{4}\text{Be} + {}^{1}_{1}\text{H}. \tag{1.2}$$

This equation does not tell us how likely the reaction is to take place, or whether it is exothermic or endothermic. It does, however, illustrate two conservation conditions that always hold: conservation of charge (Z) and conservation of nucleons $(N + Z)$. Conservation of charge requires that the sum of the subscripts on the two sides of the equation be equal, in this case $2 + 3 = 4 + 1$. Conservation of nucleons requires that the superscripts be equal, in this case $4 + 6 = 9 + 1$.

Nuclear reactions for the most part take place in two stages. First a compound nucleus is formed from the two reactants, but that nucleus is unstable and so divides, most often into two components. This being the case, we might write Eq. (1.2) in two stages:

$$ {}^{4}_{2}\text{He} + {}^{6}_{3}\text{Li} \rightarrow {}^{10}_{5}\text{B} \rightarrow {}^{9}_{4}\text{Be} + {}^{1}_{1}\text{H}. \tag{1.3}$$

However, in most of the reactions that we will utilize the compound nucleus disintegrates instantaneously. Thus no harm is done in eliminating the intermediate step from the reaction equation. The exception is when the compound nucleus is unstable but disintegrates over a longer period of time. Then, instead of writing a single equation, such as Eq. (1.3), we write two separate reaction equations. For example, when a neutron is captured by indium, it emits only a gamma ray:

$$ {}^{1}_{0}\text{n} + {}^{116}_{49}\text{In} \rightarrow {}^{117}_{49}\text{In} + {}^{0}_{0}\gamma. \tag{1.4}$$

The gamma ray has neither mass nor charge. Thus we give it both super- and subscripts of zero: ${}^{0}_{0}\gamma$. Indium-117 is not a stable nuclide but

rather undergoes radioactive decay, in this case the indium decays to
tin by emitting an electron, and an accompanying gamma ray:

$$^{117}_{49}\text{In} \rightarrow {}^{117}_{50}\text{Sn} + {}^{0}_{-1}\text{e} + {}^{0}_{0}\gamma. \tag{1.5}$$

The electron is noted by ${}^{0}_{-1}\text{e}$, with a subscript of -1, since is has the
opposite charge of a proton and a superscript of zero since its mass is
only slightly more than one two-thousandths of the mass of a proton
or neutron. A rudimentary way of looking at the nuclear model would
be to view the electron emission as resulting from one of the neutrons
within the nucleus decomposing into a proton and an electron.

Decay reactions such as Eq. (1.5) take place over time and are
characterized by a half-life, referred to as $t_{1/2}$. Given a large number
of such nuclei, half of them will decay in a time span of $t_{1/2}$, three-
fourths of them in $2t_{1/2}$, seven-eighths of them in $3t_{1/2}$, and so on.
The half-life of indium-117 is 54 minutes. Half-lives vary over many
orders of magnitude, depending on the nuclide in question. Some
radioactive materials with very long half-lives appear naturally in
the surface of the earth. For example,

$$^{234}_{92}\text{U} \rightarrow {}^{230}_{90}\text{Th} + {}^{4}_{2}\text{He} \tag{1.6}$$

with $t_{1/2} = 2.45 \times 10^5$ years. We will return to the mathematical
description of half-lives and radioactive decay later in the chapter.

Gamma rays are sometimes omitted from reaction equations;
since they carry neither mass nor charge they do not affect the nucleon
and charge balances that we have thus far discussed. Gamma rays,
however, are important in the energy conservation law that we will
discuss subsequently. Their role may be understood as follows. Following a nuclear collision, reaction, or radioactive decay the nucleus
generally is left in an excited state. It then relaxes to its ground or
unexcited state by emitting one or more gamma rays. These rays are
emitted at distinct energies, corresponding to the quantum energy
levels of the nucleus. This nuclear phenomenon is analogous to the
situation in atomic physics where an orbital electron in an excited
state emits a photon as it drops to its ground state. Both gamma rays
and photons are electromagnetic radiation. However, they differ
greatly in energy. For while the photons emitted from the relaxation
of orbital electrons typically are in the electron volt range, the energies
of gamma rays are measured in millions of electron volts.

One remaining nuclear radiation, which we have not mentioned, is
the neutrino. In conjunction with electron emission a neutrino is created, and carries off a part of the reaction energy. Since neutrinos do not
interact with matter to any significant extent, the energy they carry away
is for all practical purposes lost. However, they must be included in the
energy conservation considerations of the following section.

Notation

Before proceeding, the introduction of some shorthand notation is useful. Note from Eqs. (1.5) and (1.6) that the helium nucleus and the electron are both emitted from the decay of radionuclides. When emitted from nuclei these are referred to as alpha and beta particles, respectively. A nearly universal convention is to simplify the notation by simply referring to them as α and β particles. In like manner since gamma rays carry neither charge nor mass, and the mass and charge of neutrons and protons are simple to remember, we refer to them simply as γ, n, and p, respectively. In summary, we will often use the simplifications:

$$\ce{^4_2He} \Rightarrow \alpha \quad \ce{^0_{-1}e} \Rightarrow \beta \quad \ce{^0_0\gamma} \Rightarrow \gamma \quad \ce{^1_0n} \Rightarrow n \quad \ce{^1_1H} \Rightarrow p. \tag{1.7}$$

Likewise the notation for two important isotopes of hydrogen, deuterium and tritium, is also simplified as $\ce{^2_1H} \Rightarrow D$ and $\ce{^3_1H} \Rightarrow T$.

Instead of using the form of Eq. (1.1) we may write reaction equations more compactly as $A(B,C)D$, where the nuclei of smaller atomic number are usually the ones placed inside the parentheses. Thus, for example,

$$\ce{^1_0n} + \ce{^14_7N} \rightarrow \ce{^14_6C} + \ce{^1_1p} \tag{1.8}$$

may be compacted to $\ce{^14_7N}(n, p)\ce{^14_6C}$ or alternately as $\ce{^14_7N} \xrightarrow{(n,p)} \ce{^14_6C}$. Likewise radioactive decay such as in Eq. (1.5) is often expressed as $\ce{^117_49In} \xrightarrow{\beta} \ce{^117_50Sn}$, where in all cases it is understood that some energy is likely to be carried away as gamma rays and neutrinos.

Energetics

Einstein's equation for the equivalence between mass and energy governs the energetics of nuclear reactions:

$$E_{total} = mc^2, \tag{1.9}$$

where E_{total}, m, and c represent the total energy of a nucleus, its mass, and the speed of light, respectively. The mass in this equation, however, depends on the particles speed relative to the speed of light:

$$m = m_0 \Big/ \sqrt{1 - (v/c)^2}, \tag{1.10}$$

where m_0 is the rest mass, or the mass of the particle when its speed $v = 0$. For situations in which $v \ll c$, we may expand the square root term in powers of $(v/c)^2$,

$$m = m_0 \left[1 + \tfrac{1}{2}(v/c)^2 + O(v/c)^4\right] \qquad (1.11)$$

and retain only the first two terms. Inserting this result into Eq. (1.9), we have

$$E_{total} = m_0 c^2 + \tfrac{1}{2} m_0 v^2. \qquad (1.12)$$

The first term on the right is the rest energy, and the second is the familiar form of the kinetic energy. The neutrons found in reactors, as well as the nuclei, will always be nonrelativistic with $v \ll c$ allowing the use of Eq. (1.12). We hereafter use E to designate kinetic energy. Thus for a nonrelativistic particle with rest mass M_X, we have

$$E = \tfrac{1}{2} M_X v^2. \qquad (1.13)$$

Some high-energy electrons, however, may travel at speeds that are a substantial fraction of the speed of light, and in these cases the relativistic equations must be used. We then must determine E_{total} from Eqs. (1.9) and (1.10) and take $E = E_{total} - m_0 c^2$. Finally gamma rays have no mass and travel at the speed of light. Their energy is given by

$$E = h\nu \qquad (1.14)$$

where h is Plank's constant and ν is their frequency.

We are now prepared to apply the law that total energy must be conserved. For the reaction of Eq. (1.1) this is expressed as

$$E_A + M_A c^2 + E_B + M_B c^2 = E_C + M_C c^2 + E_D + M_D c^2, \qquad (1.15)$$

where E_A and M_A are the kinetic energy and rest mass of A, and likewise for B, C, and D. If one of the reactants is a gamma ray, then for it $E + Mc^2$ is replaced by $h\nu$ since it carries no mass. The Q of a reaction, defined as

$$Q = E_C + E_D - E_A - E_B, \qquad (1.16)$$

determines whether the reaction is exothermic or endothermic. With a positive Q kinetic energy is created, and with negative Q it is lost. Equation (1.15) allows us to write Q in terms of the masses as

$$Q = (M_A + M_B - M_C - M_D)c^2. \qquad (1.17)$$

A positive Q indicates an exothermic reaction, which creates kinetic energy and results in a net loss of rest mass. Conversely, endothermic reactions result in a net increase in rest mass. Strictly speaking these same arguments apply to chemical as to nuclear reactions. However, when one is dealing with the energy changes of the order of a few eV in chemical reactions, as opposed to changes of MeV magnitudes in nuclear reactions, the changes in mass are much too small to measure.

1.3 The Curve of Binding Energy

The foregoing conservation arguments do not indicate which nuclear reactions are likely to be exothermic or endothermic. We must examine mass defects and binding energies to understand which nuclear reactions produce rather than absorb energy. If we add the masses of the Z protons and N neutrons that make up a nucleus, say of element X, we find that the weights of these constituent masses exceed the weight M_X of the nucleus as a whole. The difference is defined as the mass defect:

$$\Delta = Z M_P + N M_N - M_X, \tag{1.18}$$

which is positive for all nuclei. Thus the nucleus weighs less than the neutrons and protons from which it is composed. Multiplying the mass defect by the square of the speed of light then yields units of energy: Δc^2. This is the binding energy of the nucleus. We may interpret it as follows. If the nucleus could be pulled apart and separated into its constituent protons and neutrons, there would be an increase in mass by an amount equal to the mass defect. Thus an equivalent amount of energy—the binding energy—would need to be expended to carry out this disassembly. All stable nuclei have positive binding energies holding them together. If we normalize the binding energy to the number of nucleons, we have

$$\Delta c^2 / (N + Z). \tag{1.19}$$

This quantity—the binding energy per nucleon—provides a measure of nuclear stability; the larger it is the more stable the nucleus will be.

Figure 1.1 is the curve of binding energy per nucleon. At low atomic mass the curve rises rapidly. For larger atomic weights, above 40 or so, the curve becomes quite smooth reaching a maximum of slightly less than 9 MeV and then gradually decreases. Exothermic reactions are those in which result in reaction products with increased binding energy, going from less to more stable nuclei. Two classes of such reaction are candidates for energy production: fusion reactions in which two light weight nuclei combine to form

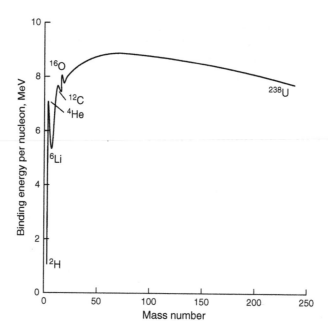

FIGURE 1.1   Curve of binding energy per nucleon.

a heaver nuclei, higher on the binding energy curve, and fission reactions in which a heavy nucleus splits to form two lighter nuclei, each with a higher binding energy per nucleon.

1.4   Fusion Reactions

Equation (1.2) is an example of a charged particle reaction, since both nuclei on the left have atomic numbers greater than zero. Such reactions are difficult to bring about, for after the orbiting electrons are stripped from the nuclei, the positive charges on the nuclei strongly repel one another. Thus to bring about a reaction such as Eq. (1.2), the nuclei must collide at high speed in order to overpower the coulomb repulsion and make contact. The most common methods for achieving such reactions on earth consist of using particle accelerations to impart a great deal of kinetic energy to one of the particles and then slam it into a target made of the second material. An alternative is to mix the two species and bring them to a very high temperature, where they become a plasma. Since the average kinetic energy of a nucleus is proportional to its absolute temperature, if high enough temperatures are reached the electrical repulsion of the nuclei is overpowered by the kinetic energy, and a thermonuclear reaction results.

Two reactions based on fusing isotopes of hydrogen have been widely considered as a basis for energy production, deuterium–deuterium and deuterium–tritium:

$$D\text{-}D \quad {}_1^2\text{H} + {}_1^2\text{H} \rightarrow {}_2^3\text{He} + {}_0^1\text{n} + 3.25\,\text{MeV},$$

$$\quad\quad {}_1^2\text{H} + {}_1^2\text{H} \rightarrow {}_1^3\text{H} + {}_1^1\text{H} + 4.02\,\text{MeV}, \quad\quad (1.20)$$

$$D\text{-}T \quad {}_1^2\text{H} + {}_1^3\text{H} \rightarrow {}_2^4\text{He} + {}_0^1\text{n} + 17.59\,\text{MeV}.$$

The difficulty is that these are charged particle reactions. Thus for the nuclei to interact the particles must be brought together with very high kinetic energies in order to overcome the coulomb repulsion of the positively charged nuclei. As a practical matter, this cannot be accomplished using a particle accelerator, for the accelerator would use much more energy than would be produced by the reaction. Rather, means must be found to achieve temperatures comparable to those found in the interior of the sun. For then the particles' heightened kinetic energy would overcome the coulomb barrier and thermonuclear reactions would result. While thermonuclear reactions are commonplace in the interior of stars, on earth the necessary temperatures have been obtained to date only in thermonuclear explosions and not in the controlled manner that would be needed for sustained power production.

Long-term efforts continue to achieve controlled temperatures high enough to obtain power from fusion reactions. Investigators place most emphasis on the D-T reaction because it becomes feasible at lower temperatures than the D-D reaction. The D-T reaction, however, has the disadvantage that most of the energy release appears as the kinetic energy of 14-MeV neutrons, which damage whatever material they impact and cause it to become radioactive.

We will not consider fusion energy further here. Rather, we will proceed to fission reactions, in which energy is released by splitting a heavy nucleus into two lighter ones that have greater binding energies per nucleon. Neutrons may initiate fission. Thus there is no requirement for high temperatures, since there is no electrical repulsion between the neutron and the nucleus. Figuratively speaking, the neutron may slide into the nucleus without coulomb resistance.

1.5 Fission Reactions

Consider now a fission reaction for uranium-235 as shown in Fig. 1.2. From the reaction come approximately 200 MeV of energy, two or three neutrons, two lighter nuclei (called fission fragments), and a number of gamma rays and neutrinos. The fission fragments undergo radioactive decay producing additional fission products. The energy

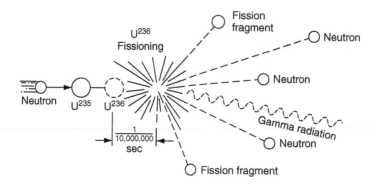

FIGURE 1.2   A fission reaction.

produced from fission, the neutrons, and the fission products all play critical roles in the physics of nuclear power reactors. We consider each of them in turn.

Energy Release and Dissipation

The approximately 200 MeV of energy released by a fission reaction appears as kinetic energy of the fission fragments, neutrons, and gamma rays, as well as that from the beta particles, gamma rays, and neutrinos emitted as the fission products undergo radioactive decay. This kinetic energy is dissipated to heat nearly instantaneously as the reaction products interact with the surrounding media. The forms that the interactions take, however, differ significantly according to whether the particles are electrically charged or neutral.

The fission fragments are highly charged, for the high speeds at which they emerge from fission cause electrons to be ripped from their shells as they encounter surrounding atoms. Charged particles interact strongly with the surrounding atoms or molecules traveling at high speed, causing them to ionize. Creation of ion pairs requires energy, which is lost from the kinetic energy of the charged particle causing it to decelerate and ultimately come to rest. The positive ions and free electrons created by the passage of the charged particle will subsequently reunite, liberating energy in the form of heat. The distance required to bring the particle to rest is referred to as its range. The range of fission fragments in solids amounts to only a few microns, and thus most of the energy of fission is converted to heat very close to the point of fission. Other charged particles, such as the alpha and beta particles emitted in radioactive decay, behave analogously, rapidly decelerating and coming to rest; for lighter charged particles the ranges are somewhat longer.

Neutrons, gamma rays, and neutrinos are neutral and behave quite differently. They are affected neither by the negative charge of electrons surrounding a nucleus nor the electric field caused by a positively charged nucleus. They thus travel in straight lines until making a collision, at which point they scatter or are absorbed. If absorbed, they cease to exist, with their energy dissipated by the collision. If they scatter, they change direction and energy, and continue along another straight line. The flight paths between collisions amount to very large numbers of interatomic distances. With neutrinos these distances are nearly infinite; for neutrons and gamma rays traveling in solids they are typically measured in centimeters. Neutrons scatter only from nuclei, whereas gamma rays are scattered by electrons as well. Except at very low energies, a neutron will impart significant kinetic energy to the nucleus, causing it to become striped of orbital electrons and therefore charged. The electrons that gain kinetic energy from gamma ray collisions, of course, are already charged. In either case the collision partner will decelerate and come to rest in distances measured in microns, dissipating its energy as heat very close to the collision site.

More than 80% of the energy released by fission appears as the kinetic energy of the fission fragments. The neutrons, beta particles, and gamma and neutrino radiation account for the remainder. The energy of the neutrinos, however, is lost because they travel nearly infinite distances without interacting with matter. The remainder of the energy is recovered as heat within a reactor. This varies slightly between fissionable isotopes; for uranium-235 it is approximately 193 MeV or $\gamma = 3.1 \times 10^{-11}$ J/fission.

The difference in energy dissipation mechanisms between charged and neutral particles also causes them to create biological hazards by quite different mechanisms. The alpha and beta radiation emitted by fission products or other radioisotopes are charged particles. They are referred to as nonpenetrating radiation since they deposit their energy over a very short distance or range. Alpha or beta radiation will not penetrate the skin and therefore is not a significant hazard if the source is external to the body. They pose more serious problems if radioisotopes emitting them are inhaled or ingested. Then they can attack the lungs and digestive tract, and other organs as well, depending on the biochemical properties of the radioisotope. Radiostrontium, for example, collects in the bone marrow and does its damage there, whereas for radioiodine the thyroid gland is the critical organ. In contrast, since neutral particles (neutrons and gamma rays) travel distances measured in centimeters between collisions in tissue, they are primarily a hazard from external sources. The damage neutral particles do is more uniformly distributed over the whole body, resulting from the ionization of water and other tissue molecules at the points where neutrons collide with nuclei or gamma rays with electrons.

Neutron Multiplication

The two or three neutrons born with each fission undergo a number of scattering collisions with nuclei before ending their lives in absorption collisions, which in many cases cause the absorbing nucleus to become radioactive. If the neutron is absorbed in a fissionable material, frequently it will cause the nucleus to fission and give birth to neutrons of the next generation. Since this process may then be repeated to create successive generations of neutrons, a neutron chain reaction is said to exist. We characterize this process by defining the chain reaction's multiplication, k, as the ratio of fission neutrons born in one generation to those born in the preceding generation. For purposes of analysis, we also define a neutron lifetime in such a situation as beginning with neutron emission from fission, progressing—or we might say aging—though a succession of scattering collisions, and ending with absorption.

Suppose at some time, say $t = 0$, we have n_o neutrons produced by fission; we shall call these the zeroth generation. Then the first generation will contain kn_o neutrons, the second generation $k^2 n_o$, and so on: the ith generation will contain $k^i n_o$. On average, the time at which the ith generation is born will be $t = i \cdot l$, where l is the neutron lifetime. We can eliminate i between these expressions to estimate the number of neutrons present at time t:

$$n(t) = n_o k^{t/l}. \tag{1.21}$$

Thus the neutron population will increase, decrease, or remain the same according to whether k is greater than, less than, or equal to one. The system is then said to be supercritical, subcritical, or critical, respectively.

A more widely used form of Eq. (1.21) results if we limit our attention to situations where k is close to one: First note that the exponential and natural logarithm are inverse functions. Thus for any quantity, say x, we can write $x = \exp[\ln(x)]$ Thus with $x = k^{t/l}$ we may write Eq. (1.21) as

$$n(t) = n_o \exp[(t/l) \ln(k)]. \tag{1.22}$$

If k is close to one, that is, $|k - 1| \ll 1$, we may expand $\ln(k)$ about 1 as $\ln(k) \approx k - 1$, to yield:

$$n(t) = n_o \exp[(k - 1)t/l]. \tag{1.23}$$

Thus the progeny of the neutrons created at time zero behaves exponentially as indicated in Fig. 1.3. Much of the content of the

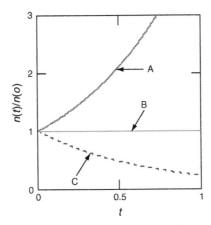

FIGURE 1.3   Neutron population versus time in (A) supercritical system, (B) critical system, (C) subcritical system.

following chapters deals with the determination of the multiplication, how it depends on the composition and size of a reactor, and how the time-dependent behavior of a chain reaction is affected by the presence of the small fraction of neutrons whose emission following fission is delayed. Subsequently we will examine changes in multiplication caused by changes in temperature, fuel depletion, and other factors central to the design and operation of power reactors.

Fission Products

Fission results in many different pairs of fission fragments. In most cases one has a substantially heavier mass than the other. For example, a typical fission reaction is

$$n + {}^{235}_{92}U \rightarrow {}^{140}_{54}Xe + {}^{94}_{38}Sr + 2n + 200\,\text{MeV}. \tag{1.24}$$

Fission fragments are unstable because they have neutron to proton ratios that are too large. Figure 1.4, which plots neutrons versus protons, indicates an upward curvature in the line of stable nuclei, indicating that the ratio of neutrons to protons increases above 1:1 as the atomic number becomes larger (e.g., the prominent isotopes of carbon and oxygen are ${}^{12}_{6}C$ and ${}^{16}_{8}O$ but for lead and thorium they are ${}^{207}_{82}Pb$ and ${}^{232}_{90}Th$). As a nucleus fissions the ratio of neutrons to protons would stay the same in the fission fragments—as indicated by the dashed line in Fig. 1.4—were it not for the 2 to 3 neutrons given off promptly at the time of fission. Even then, the fission fragments lie above the curve of stable nuclei. Less than 1%

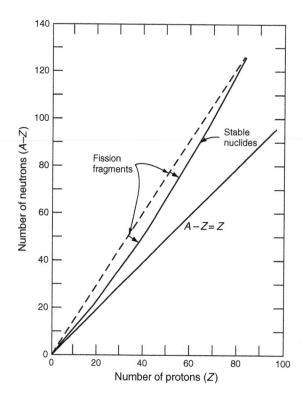

FIGURE 1.4   Fission fragment instability.

of these fragments decay by the delayed emission of neutrons. The predominate decay mode is through beta emission, accompanied by one or more gamma rays. Such decay moves the resulting nuclide toward the line of stable nuclei as the arrows in Fig. 1.4 indicate. However, more than one decay is most often required to arrive at the range of stable nuclei. For the fission fragments in Eq. (1.24) we have

$$\,^{140}_{54}\text{Xe} \xrightarrow{\beta} \,^{140}_{55}\text{Cs} \xrightarrow{\beta} \,^{140}_{56}\text{Ba} \xrightarrow{\beta} \,^{140}_{57}\text{La} \xrightarrow{\beta} \,^{140}_{58}\text{Ce} \qquad (1.25)$$

and

$$\,^{94}_{38}\text{Sr} \xrightarrow{\beta} \,^{94}_{39}\text{Y} \xrightarrow{\beta} \,^{94}_{40}\text{Zr}. \qquad (1.26)$$

Each of these decays has a characteristic half-life. With some notable exceptions the half-lives earlier in the decay chain tend to be shorter than those occurring later. The fission fragments taken together with their decay products are classified as fission products.

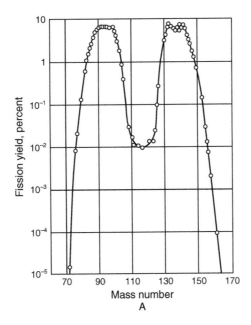

FIGURE 1.5   Fission product yields for uranium-235.

Equation (1.24) shows only one example of the more than 40 different fragment pairs that result from fission. Fission fragments have atomic mass numbers between 72 and 160. Figure 1.5 shows the mass frequency distribution for uranium-235, which is typical for other fissionable materials provided the neutrons causing fission have energies of a few eV or less. Nearly all of the fission products fall into two broad groups. The light group has mass numbers between 80 and 110, whereas the heavy group has mass numbers between 125 and 155. The probability of fissions yielding products of equal mass increases with the energy of the incident neutron, and the valley in the curve nearly disappears for fissions caused by neutrons with energies in the tens of MeV. Because virtually all of the 40 fission product pairs produce characteristic chains of radioactive decay from successive beta emissions, more than 200 different radioactive fission products are produced in a nuclear reactor.

Roughly 8% of the 200 MeV of energy produced from fission is attributable to the beta decay of fission products and the gamma rays associated with it. Thus even following shutdown of a chain reaction, radioactive decay will continue to produce significant amounts of heat. Figure 1.6 shows the decay heat for a reactor that has operated at a power P for a long time. The heat is approximated by the Wigner-Way formula as

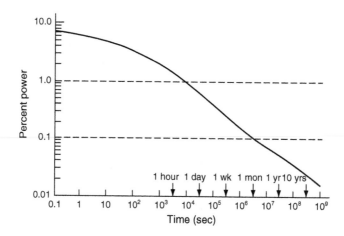

FIGURE 1.6   Heat produced by decay of fission products.

$$P_d(t) = 0.0622\, P_o\left[t^{-0.2} - (t_o + t)^{-0.2}\right] \tag{1.27}$$

where

$P_d(t)$ = power generation due to beta and gamma rays,
  P_o = power before shutdown,
  t_o = time, in seconds, of power operation before shutdown,
   t = time, in seconds, elapsed since shutdown.

As a result of decay heat, cooling must be provided to prevent over-heating of reactor fuel for a substantial period of time following power plant shutdown.

1.6   Fissile and Fertile Materials

In discussing nuclear reactors we must distinguish between two classes of fissionable materials. A fissile material is one that will undergo fission when bombarded by neutrons of any energy. The isotope uranium-235 is a fissile material. A fertile material is one that will capture a neutron, and transmute by radioactive decay into a fissile material. Uranium-238 is a fertile material. Fertile isotopes may also undergo fission directly, but only if impacted by a high-energy neutron, typically in the MeV range. Thus fissile and fertile materials together are defined as fissionable materials. Fertile materials by themselves, however, are not capable of sustaining a chain reaction.

Uranium-235 is the only naturally occurring fissile material. Moreover, it constitutes only 0.7% of natural uranium. Except for trace amounts of other isotopes, uranium-238 constitutes the remaining 99.3% of natural uranium. By capturing a neutron, uranium-238 becomes radioactive and decays to plutonium-239:

$$n + {}^{238}_{92}U \longrightarrow {}^{239}_{92}U \xrightarrow{\beta} {}^{239}_{93}Np \xrightarrow{\beta} {}^{239}_{94}Pu. \qquad (1.28)$$

If a neutron of any energy strikes plutonium-239, there is a strong probability that it will cause fission. Thus it is a fissile isotope. Plutonium-239 itself is radioactive. However its half-life of 24.4 thousand years is plenty long enough that it can be stored and used as a reactor fuel. There is a smaller probability that the plutonium will simply capture the neutron, resulting in the reaction

$$n + {}^{239}_{94}Pu \rightarrow {}^{240}_{94}Pu. \qquad (1.29)$$

Plutonium-240, however, is again a fertile material. If it captures a second neutron it will become plutonium-241, a fissile material.

In addition to uranium-238, a second fertile material occurring in nature is thorium-232. Upon capturing a neutron it undergoes decay as follows:

$$n + {}^{232}_{90}Th \longrightarrow {}^{233}_{90}Th \xrightarrow{\beta} {}^{233}_{91}Pa \xrightarrow{\beta} {}^{233}_{92}U, \qquad (1.30)$$

yielding the fissile material uranium-233. This reaction is of particular interest for sustaining nuclear energy over the very long term since the earth's crust contains substantially more thorium than uranium.

Fissile materials can be produced by including the parent fertile material in a reactor core. Returning to Fig. 1.2, we see that if more than two neutrons are produced per fission—and the number is about 2.4 for uranium-235—then there is the possibility of utilizing one neutron to sustain the chain reaction, and more than one to convert fertile to fissile material. If this process creates more fissile material than it destroys, the reactor is said to be a breeder; it breeds more fissile material than it consumes.

Since most power reactors are fueled by natural or partially enriched uranium, there is a bountiful supply of uranium-238 in the reactor for conversion to plutonium. However, as subsequent chapters will detail, to sustain breeding the designer must prevent a large fraction of the fission neutrons from being absorbed in nonfissile materials or from leaking from the reactor. This is a major challenge. Most reactors burn more fissile material than they create.

Because half-lives, cross sections, and other properties of fissile and fertile isotopes are ubiquitous to reactor theory, the following unambiguous shorthand frequently is used for their designation.

Their properties are designated by the last digits of their atomic charge, and atomic mass: Thus properties of fissionable element $^{abc}_{de}X$ are simply designated sub- or superscripts "*ec.*" For example, $^{232}_{90}Th \rightarrow 02$, $^{235}_{92}U \rightarrow 25$, $^{238}_{92}U \rightarrow 28$, and $^{239}_{94}Pu \rightarrow 49$.

One question remains: Where do the neutrons come from to initiate a chain reaction? Some neutrons occur naturally, as the result of very high-energy cosmic rays colliding with nuclei and causing neutrons to be ejected. If no other source were present these would trigger a chain reaction. Invariably, a stronger and more reliable source is desirable. Although there are a number of possibilities, probably the most widely used is the radium beryllium source. It combines the alpha decay of a naturally occurring radium isotope

$$^{226}_{88}Ra \xrightarrow{\alpha} {}^{222}_{86}Rn, \tag{1.31}$$

which has a half-life of 1600 years with the reaction

$$^{9}_{4}Be \xrightarrow{(\alpha,n)} {}^{12}_{6}C \tag{1.32}$$

to provide the needed neutrons.

1.7 Radioactive Decay

To understand the behavior of fission products, the rates of conversion of fertile to fissile materials, and a number of other phenomena related to reactor physics we must quantify the behavior of radioactive materials. The law governing the decay of a nucleus states that the rate of decay is proportional to the number of nuclei present. Each radioisotope—that is, an isotope that undergoes radioactive decay—has a characteristic decay constant λ. Thus if the number of nuclei present at time t is $N(t)$, the rate at which they decay is

$$\frac{d}{dt}N(t) = -\lambda N(t). \tag{1.33}$$

Dividing by $N(t)$, we may integrate this equation from time zero to t, to obtain

$$\int_{N(0)}^{N(t)} dN/N = -\lambda \int_{0}^{t} dt, \tag{1.34}$$

where $N(0)$ is the initial number of nuclei. Noting that $dN/N = d\ln(N)$, Eq. (1.34) becomes

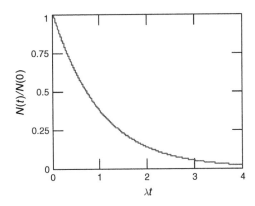

FIGURE 1.7   Exponential decay of a radionuclide.

$$\ln[N(t)/N(0)] = -\lambda t, \tag{1.35}$$

yielding the characteristic exponential rate of decay,

$$N(t) = N(0)\exp(-\lambda t). \tag{1.36}$$

Figure 1.7 illustrates the exponential decay of a radioactive material.

The half-life, $t_{1/2}$, is a more intuitive measure of the times over which unstable nuclei decay. As defined earlier, $t_{1/2}$ is the length of time required for one-half of the nuclei to decay. Thus it may be obtained by substituting $N(t_{1/2}) = N(0)/2$ into Eq. (1.35) to yield $\ln(1/2) = -0.693 = -\lambda t_{1/2}$, or simply

$$t_{1/2} = 0.693/\lambda. \tag{1.37}$$

A second, less-used measure of decay time is the mean time to decay, defined by

$$\bar{t} = \int_0^\infty tN(t)dt \bigg/ \int_0^\infty N(t)dt = 1/\lambda. \tag{1.38}$$

Before proceeding, a word is in order concerning units. Normally we specify the strength of a radioactive source in terms of curies (Ci) where 1 Ci is defined as 3.7×10^{10} disintegrations per second, which is the rate decay of one gram of radium-226; the becquerel (Bq), defined as one disintegration per second, has also come into use as a measure of radioactivity. To calculate the number of nuclei present we first note that Avogadro's number,

$N_o = 0.6023 \cdot 10^{24}$, is the number of atoms in one gram molecular weight, and thus the total number of atoms is just mN_o/A where m is the mass in grams and A is the atomic mass of the isotope. The concentration in atoms/cm^3 is then $\rho N_o/A$, where ρ is the density in grams/cm^3.

Saturation Activity

Radionuclides are produced at a constant rate in a number of situations. For example, a reactor operating at constant power produces radioactive fission fragments at a constant rate. In such situations, we determine the time dependence of the inventory of an isotope produced at a rate of A_o nuclei per unit time by adding a source term A_o to Eq. (1.33):

$$\frac{d}{dt} N(t) = A_o - \lambda N(t). \tag{1.39}$$

To solve this equation, multiply both sides by an integrating factor of $\exp(\lambda t)$. Then utilizing the fact that

$$\frac{d}{dt}[N(t)\exp(\lambda t)] = \left[\frac{d}{dt}N(t) + \lambda N(t)\right]\exp(\lambda t), \tag{1.40}$$

we have

$$\frac{d}{dt}[N(t)\exp(\lambda t)] = A_o \exp(\lambda t). \tag{1.41}$$

Now if we assume that initially there are no radionuclides present, that is, $N(0) = 0$, we may integrate this equation between 0 and t and obtain

$$\lambda N(t) = A_o[1 - \exp(-\lambda t)], \tag{1.42}$$

where $\lambda N(t)$ is the activity measured in disintegrations per unit time. Note that initially the activity increases linearly with time, since for $\lambda t \ll 1$, $\exp(-\lambda t) \approx 1 - \lambda t$. After several half-lives, however, the exponential term becomes vanishingly small, and the rate of decay is then equal to the rate of production or $\lambda N(\infty) = A_o$. This is referred to as the saturation activity. Figure 1.8 illustrates the buildup to saturation activity given by Eq. (1.42).

To illustrate the importance of saturation activity, consider iodine-131 and strontium-90, which are two of the more important fission products stemming from the operation of power reactors. Assume a power reactor produces them at rates of $0.85 \cdot 10^{18}$ nuclei/s

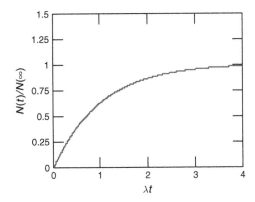

FIGURE 1.8 Activity versus time for a radionuclide produced at a constant rate.

and $1.63 \cdot 10^{18}$ nuclei/s, respectively, and ask how many curies of activity each produces after 1 week, 1 month, and 1 year of operation.

The two isotopes have half-lives of 8.05 days and 10,628 days. Thus from Eq. (1.37) we have $\lambda_I = 0.0861$/day, and $\lambda_{Sr} = 6.52 \cdot 10^{-5}$/day. To express the activity in curies we divide Eq. (1.42) by $3.7 \cdot 10^{10}$ nuclei/s. Thus $A_I = 2.30 \cdot 10^7$Ci, and $A_{Sr} = 4.40 \cdot 10^7$Ci. We take $t = 7$ days, 30 days, and 365 days (i.e., 1 week, 1 month, and 1 year) in Eq. (1.42) and obtain:

$$\lambda_I N_I(7) = 10.4 \cdot 10^6 \text{ Ci}, \qquad \lambda_{Sr} N_{Sr}(7) = 2.01 \cdot 10^3 \text{ Ci}$$
$$\lambda_I N_I(30) = 21.2 \cdot 10^6 \text{ Ci}, \qquad \lambda_{Sr} N_{Sr}(30) = 8.61 \cdot 10^4 \text{ Ci}$$
$$\lambda_I N_I(365.25) = 23.0 \cdot 10^6 \text{ Ci}, \qquad \lambda_{Sr} N_{Sr}(365.25) = 1.04 \cdot 10^6 \text{ Ci}.$$

The shorter half-lived iodine-131 has nearly reached saturation at the end of 1 month, and remains constant thereafter with a value that is proportional to the reactor power. In contrast the activity of strontium-90, with a much longer half-life, increases linearly with time and will continue to do so for a number of years. The plot of activity versus λt shown in Fig. 1.8 illustrates these effects more clearly. At $t = 1$ year, $\lambda_{Sr} t = 6.52 \cdot 10^{-5} \cdot 365.25 = 0.0238 \ll 1$, which is far short of the time required to reach saturation. Thus over the first year—and for substantially longer—the inventory of strontium-90 will grow in proportion to the total energy that the reactor has produced since start-up. In contrast, at 1 month $\lambda_I t = 0.0861 \cdot 30 = 2.58$ and thus, as Fig. 1.8 indicates, iodine-131 is very close to saturation.

Decay Chains

The foregoing reactions may be represented as a simple decay process: $A \rightarrow B + C$. As Eqs. (1.25) and (1.26) indicate, however, chains of decay often occur. Consider the two-stage decay

$$A \rightarrow B + C$$
$$\searrow \qquad (1.43)$$
$$D + E$$

and let the decay constants of A and B be denoted by λ_A and λ_B. For isotope A we already have the solution in the form of Eq. (1.36). Adding subscripts to distinguish it from B, we have

$$N_A(t) = N_A(0) \exp(-\lambda_A t), \qquad (1.44)$$

and $\lambda_A N_A(t)$ is the number of nuclei of type A decaying per unit time. Since for each decay of a nucleus of type A a nucleus of type B is produced, the rate at which nuclei of type B is produced is also $\lambda_A N_A(t)$. Likewise if there are $N_B(t)$ of isotope B present, its rate of decay will be $\lambda_B N_B(t)$. Thus the net rate of creation of isotope B is

$$\frac{d}{dt} N_B(t) = \lambda_A N_A(t) - \lambda_B N_B(t). \qquad (1.45)$$

To solve this equation, we first replace $N_A(t)$ by Eq. (1.44). We then move $\lambda_B N_B(t)$ to the left and use the same integrating factor technique as before: We multiply both sides of the equation by $\exp(\lambda_B t)$ and employ Eq. (1.40) to simplify the left-hand side:

$$\frac{d}{dt} [N_B(t) \exp(\lambda_B t)] = \lambda_A N_A(0) \exp[(\lambda_B - \lambda_A)t]. \qquad (1.46)$$

Multiplying by dt and then integrating from 0 to t yields

$$N_B(t) \exp(\lambda_B t) - N_B(0) = \frac{\lambda_A}{\lambda_B - \lambda_A} N_A(0)\{\exp[(\lambda_B - \lambda_A)t] - 1\}. \quad (1.47)$$

If we assume that the isotope B is not present initially so that $N_B(0) = 0$, we have

$$N_B(t) = \frac{\lambda_A}{\lambda_B - \lambda_A} N_A(0)(e^{-\lambda_A t} - e^{-\lambda_B t}). \qquad (1.48)$$

Figure 1.9 shows the time-dependent behavior of the activities $A_A(t) = \lambda_A N_A(t)$ and $A_B(t) = \lambda_B N_B(t)$ for cases for which $\lambda_A \ll \lambda_B$, $\lambda_A \gg \lambda_B$, and $\lambda_A \cong \lambda_B$. If $\lambda_A \ll \lambda_B$, that is, if the half-life of A is much longer than that of B, then $\exp(-\lambda_B t)$ decays much faster than $\exp(-\lambda_A t)$ and after a few half-lives of B we obtain from Eqs. (1.44) and (1.48) $\lambda_B N_B(t) \approx \lambda_A N_A(t)$, meaning that the decay rates of A and B are approximately equal. This is referred to as secular equilibrium.

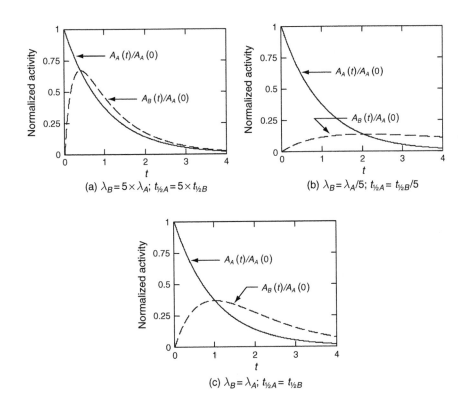

FIGURE 1.9   Decay of a sequence of two radionuclides.

On the other hand, if $\lambda_A \gg \lambda_B$, that is, if the half-life of A is much shorter than that of B, then $\exp(-\lambda_A t)$ will decay much faster than $\exp(-\lambda_B t)$, and after a few half-lives of A we can assume that it has vanished. In that case Eq. (1.48) reduces to $N_B(t) \approx N_A(0) \exp(-\lambda_B t)$. Of course, if $\lambda_A \cong \lambda_B$, neither of these approximations hold.

Bibliography

Bodansky, David, *Nuclear Energy: Principles, Procedures, and Prospects,* Springer, 2004.

Cember, H., *Introduction to Health Physics,* 3rd ed., McGraw-Hill, NY, 1996.

Duderstadt, James J., and Louis J. Hamilton, *Nuclear Reactor Analysis,* Wiley, NY, 1976.

Glasstone, Samuel, and Alexander Sesonske, *Nuclear Reactor Engineering,* 3rd ed., Van Nostrand-Reinhold, NY, 1981.

Knief, Ronald A., *Nuclear Energy Technology: Theory and Practice of Commercial Nuclear Power,* McGraw-Hill, NY, 1981.

Lamarsh, John R., *Introduction to Nuclear Reactor Theory*, Addison-Wesley, Reading, MA, 1972.

Lamarsh, John, and Anthony J. Baratta, *Introduction to Nuclear Engineering*, 3rd ed., Prentice-Hall, Englewood, NJ, 2001.

Stacey, Weston M., *Nuclear Reactor Physics*, Wiley, NY, 2001.

Williams, W. S. C., *Nuclear and Particle Physics*, Oxford University Press, USA, NY, 1991.

Wong, Samuel M., *Introductory Nuclear Physics*, 2nd ed., Wiley, NY, 1999. http://www.webelements.com/webelements/scholar/

Problems

1.1. The following isotopes frequently appear in reactor cores. What are their chemical symbols and names?

a. $^{90}_{38}?$ b. $^{91}_{40}?$ c. $^{137}_{55}?$ d. $^{157}_{64}?$ e. $^{178}_{72}?$ f. $^{137}_{93}?$ g. $^{241}_{95}?$

1.2. There are several possible modes of disintegration for the unstable nucleus $^{27}_{13}$Al. Complete the following reactions: $^{27}_{13}$Al $\rightarrow$? $+\, ^{1}_{0}$n, $^{27}_{13}$Al $\rightarrow$? $+\, ^{1}_{1}$p, $^{27}_{13}$Al $\rightarrow$? $+\, ^{2}_{1}$H, $^{27}_{13}$Al $\rightarrow$? $+\, ^{4}_{2}$He

1.3. Complete the following reactions: $^{9}_{?}$Be $+\, ^{4}_{2}$He $\rightarrow$? $+\, ^{1}_{1}$H, $^{60}_{?}$Co $\rightarrow$? $+\, ^{0}_{-1}$e, $^{7}_{3}$Li $+\, ^{1}_{1}$H $\rightarrow$? $+\, ^{4}_{2}$He, $^{10}_{5}$B $+\, ^{4}_{2}$He $\rightarrow$? $+\, ^{1}_{1}$H

1.4. What target isotope must be used for forming the compound nucleus $^{60}_{28}$Ni if the incident projective is

a. an alpha particle
b. a proton
c. a neutron?

1.5. The average kinetic energy of a fission neutron is 2.0 MeV. Defining the kinetic energy as $E_{total} - m_0 c^2$, what is the percent error introduced into the kinetic energy from using Eq. (1.12) instead of Eq. (1.9)?

1.6. Consider the following nuclear and chemical reactions:

a. A uranium-235 nucleus fissions as a result of being bombarded by a slow neutron. If the energy of fission is 200 MeV, approximately what fraction of the reactant's mass is converted to energy?

b. A carbon-12 atom undergoes combustion following collision with an oxygen-16 molecule, forming carbon dioxide. If 4 eV of energy is released, approximately what fraction of the reactant's mass is converted to energy?

1.7. a. If plutonium-239 captures two neutrons followed by a beta decay, what isotope is produced?
   b. If plutonium-239 captures three neutrons, followed by two beta decays, what isotope is produced?

1.8. To first approximation a nucleus may be considered to be a sphere with the radius in cm given by $R = 1.25 \cdot 10^{-13} A^{1/3}$ cm, where A is the atomic mass number. What are the radii of

   a. hydrogen
   b. carbon-12
   c. xenon-140
   d. uranium-238?

1.9. A reactor operates at a power of 10^3 MW(t) for 1 year. Calculate the power from decay heat

   a. 1 day following shutdown,
   b. 1 month following shutdown,
   c. 1 year following shutdown.
   d. Repeat a, b, and c, assuming only one month of operation, and compare results.

1.10. In Eq. (1.28) the uranium-239 and neptunium-239 both undergo beta decay with half-lives of 23.4 m and 2.36 d, respectively. If neutron bombardment in a reactor causes uranium-239 to be produced at a constant rate, how long will it take plutonium-239 to reach

   a. ½ of its saturation activity
   b. 90% of its saturation activity
   c. 99% of its saturation activity? (Assume that plutonium-239 undergoes no further reactions.)

1.11. Uranium-238 has a half-life of 4.51×10^9 yr, whereas the half-life of uranium-235 is only 7.13×10^8 yr. Thus since the earth was formed 4.5 billion years ago, the isotopic abundance of uranium-235 has been steadily decreasing.

   a. What was the enrichment of uranium when the earth was formed?
   b. How long ago was the enrichment 4%?

1.12. How many curies of radium-226 are needed in the reaction given in Eqs. (1.31) and (1.32) to produce 10^6 neutrons/s?

1.13. Suppose that a specimen is placed in a reactor, and neutron bombardment causes a radioisotope to be produced at a rate of 2×10^{12} nuclei/s. The radioisotope has a half-life of 2 weeks. How long should the specimen be irradiated to produce 25 Ci of the radioisotope?

1.14. The decay constant for the radioactive antimony isotope $^{124}_{51}\mathrm{Sb}$ is $1.33 \times 10^{-7}\,\mathrm{s}^{-1}$.

   a. What is its half-life in years?
   b. How many years would it take for it to decay to 0.01% of its initial value?
   c. If it were produced at a constant rate, how many years would it take to reach 95% of its saturation value?

1.15. Approximately what mass of cobalt-60, which has a half-life of 5.26 yr, will have the same number of curies as 10 g of strontium-90, which has a half-life of 28.8 yr?

1.16. Ninety percent of an isotope decays in 3 hours.

   a. What fraction decays in 6 hours?
   b. What is the half-life?
   c. If the isotope is produced in a reactor at the rate of 10^9 nuclei per hour, after a long time how many nuclei will be present in the reactor?

1.17. A fission product A with a half-life of 2 weeks is produced at the rate of 5.0×10^8 nuclei/s in a reactor.

   a. What is the saturation activity in disintegrations per second?
   b. What is the saturation activity in curies?
   c. How long after the start-up of the reactor will 90 percent of the saturation activity be reached?
   d. If the fission product undergoes decay $A \rightarrow B \rightarrow C$, where B also has a 2-week half-life, what will be the activity of B after 2 weeks?

1.18. Suppose the radioactive cobalt and strontium sources in problem 1.15 are allowed to decay for 10 years. It is found that after 10 years 1.0 Ci of cobalt-60 remains. How many curies of strontium-90 will remain?

1.19. Polonium-210 decays to lead-206 by emitting an alpha particle with a half-life of 138 days, and an energy of 5.305 MeV.

   a. How many curies are there in 1 g of pure polonium?
   b. How many watts of heat are produced by 1 g of polonium?

1.20 Consider the fission product chain $A \xrightarrow{\beta} B \xrightarrow{\beta} C$ with decay constants λ_A and λ_B. A reactor is started up at $t=0$ and produces fission product A at a rate of A_o thereafter. Assuming that B and C are not produced directly from fission:

a. Find $N_A(t)$ and $N_B(t)$.
b. What are $N_A(\infty)$ and $N_B(\infty)$?

CHAPTER 2

Neutron Interactions

2.1 Introduction

The behavior of the neutrons emitted from fission as they interact with matter determines the nature of neutron chain reactions, for to create a sustained chain reaction, on average one of the two or more neutrons created by each fission must survive to create a subsequent fission. The kinetic energy of the neutrons as well as the manner in which they travel though space and interact with nuclei lie at the basis of their behavior in nuclear reactors. At the core of neutron interactions is the concept of the cross section—that is, the cross-sectional area of a nucleus as it appears to an oncoming neutron. Such cross sections, their dependence on the neutron's kinetic energy, and the relative probabilities that a collision will result in scattering, capture, or fission form the basic physical data upon which the properties of chain reactions rest.

This chapter first describes neutrons' behavior as they travel through space and defines microscopic and macroscopic cross sections. We then distinguish between cross sections for scattering, absorption, and other reaction types. After determining the range of kinetic energies over which neutrons may exist in a reactor, we describe the dependence of cross section on neutron energy, and then conclude the chapter by describing the distributions of energies of scattered neutrons.

2.2 Neutron Cross Sections

Neutrons are neutral particles. Neither the electrons surrounding a nucleus nor the electric field caused by a positively charged nucleus affect a neutron's flight. Thus neutrons travel in straight lines, deviating from their path only when they actually collide with a nucleus to be scattered into a new direction or absorbed. The life of a neutron thus consists typically of a number of scattering collisions followed by absorption at which time its identity is lost. To a neutron traveling through a solid, space appears to be quite empty. Since an atom has a radius typically of the order of 10^{-8} cm and a nucleus only of the order of 10^{-12} cm, the fraction of the cross-sectional area perpendicular to a neutron's flight path blocked by a single tightly packed layer of atoms would be roughly $(10^{-12})^2/(10^{-8})^2 = 10^{-8}$, a small fraction indeed. Thus neutrons on average penetrate many millions of layers of atoms between collisions with nuclei. If the target material is thin—say, a piece of paper—nearly all of neutrons would be expected to pass through it without making a collision.

Microscopic and Macroscopic Cross Sections

To examine how neutrons interact with nuclei, we consider a beam of neutrons traveling in the x direction as indicated in Fig. 2.1. If the beam contains n''' neutrons per cm^3 all traveling with a speed v in the x direction, we designate $I = n'''v$ as the beam intensity. With the speed measured in cm/s, the beam intensity units are neutrons/cm^2/s. Assume that if a neutron collides with a nucleus it will either be absorbed or be scattered into a different direction. Then only neutrons that have not collided will remain traveling in the x direction. This causes the intensity of the uncollided beam to diminish as it penetrates deeper into the material.

Let $I(x)$ represent the beam intensity after penetrating x cm into the material. In traveling an additional infinitesimal distance dx, the fraction of neutrons colliding will be the same as the fraction of

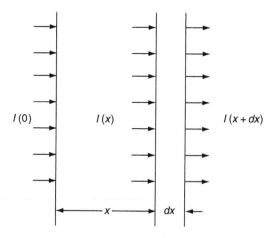

FIGURE 2.1 Neutron passage through a slab.

the 1-cm^2 section perpendicular to the beam direction that is shadowed by nuclei. If dx is small, and the nuclei are randomly placed, then the shadowing of one nucleus by another can be ignored. (Only in the rarely encountered circumstance where neutrons are passing through a single crystal does this assumption break down.) Now assume there are N nuclei/cm^3 of the material; there will then be $N\,dx$ per cm^2 in the infinitesimal thickness. If each nucleus has a cross-sectional area of σ cm^2, then the fraction of the area blocked is $N\sigma dx$, and thus we have

$$I(x + dx) = (1 - N\sigma dx)I(x). \qquad (2.1)$$

Using the definition of the derivative, we obtain the simple differential equation

$$\frac{d}{dx}I(x) = -N\sigma I(x), \qquad (2.2)$$

which may rewritten as

$$\frac{dI(x)}{I(x)} = -N\sigma dx \qquad (2.3)$$

and integrated between 0 and x to yield

$$I(x) = I(0)\exp(-N\sigma x). \qquad (2.4)$$

We next define the macroscopic cross section as

$$\Sigma = N\sigma. \qquad (2.5)$$

Here σ, which has units of cm^2/nucleus, is referred to as the *microscopic* cross section. Since the units of N are nuclei/cm^3, Σ, the *macroscopic* cross section, must have units of cm^{-1}.

The cross section of a nucleus is very small. Thus instead of measuring microscopic cross sections in cm^2 the unit of the *barn* is commonly used. One barn, abbreviated as "b," is equal to 10^{-24} cm^2. The unit is said to have originated from early determinations of neutron cross sections when one of the investigators exclaimed, "That's as big as a barn" in reaction to one of the measurements.

The foregoing equations have a probabilistic interpretation. Since $dI(x)$ is the number of neutrons that collide in dx, out of a total of $I(x)$, $-dI(x)/I(x) = \Sigma dx$, as given by Eq. (2.3), must be the probability that a neutron that has survived without colliding until x, will collide in the next dx. Likewise $I(x)/I(0) = \exp(-\Sigma x)$ is the fraction of neutrons that have moved through a distance x without colliding; it may also be interpreted as the probability of a neutron traveling a distance x without making a collision. If we then ask what is the probability $p(x)dx$ that a neutron will make its first collision in dx, it is the probability that it has survived to dx and that it will collide in dx. If its probability of colliding in dx is independent of its past history, the required result is obtained simply by multiplying the probabilities together, yielding

$$p(x)dx = \Sigma \exp(-\Sigma x)dx. \tag{2.6}$$

From this we can calculate the mean distance traveled by a neutron between collisions. It is called the mean free path and denoted by λ:

$$\lambda = \int_0^\infty xp(x)dx = \int_0^\infty x\Sigma \exp(-\Sigma x)dx = 1/\Sigma. \tag{2.7}$$

Thus the mean free path is just the inverse of the macroscopic cross section.

Uncollided Flux

The neutrons included in $I(x)$ have not made a collision. They are sometimes designated as an uncollided flux to distinguish them from the total population of neutrons, which also includes those that have made one or more collisions. The neutrons in $I(x)$ all travel in the same positive x direction, while those that have made collisions may be found traveling in all directions. The neutron beam $I(x)$ may be written as the product of the neutron speed v, in cm/s, and $n_u'''(x)$, the density of uncollided neutrons, measured in neutrons/cm^3. We thus have $I(x) = vn_u'''(x)$, and it is this form in which the flux, which is usually designated by ϕ, is written. Thus for the neutron beam the uncollided flux is

$$\phi_u(x) = vn_u'''(x). \tag{2.8}$$

The uncollided flux may be written for other configurations than the beam of neutrons used here to define the cross section. A point source is particularly useful in distinguishing the difference between geometric and material attenuation of the uncollided flux. Let a source emit s_p neutrons per second. At any location all of the uncollided neutrons travel in a single direction: radially outward from the source. In a vacuum the flux is only attenuated geometrically since no material is present: at a distance r from the source the neutrons will pass through a surface of area $4\pi r^2$ of a sphere of radius r, and thus the number passing through 1 cm^2/s is $\phi_u(r) = s_p/(4\pi r^2)$. With a material present, however, only a fraction $\exp(-\Sigma r)$ of the neutrons will survive to a distance r without undergoing a collision. Thus accounting for both geometrical and material attenuations results in the uncollided flux at a distance r from a point source being

$$\phi_u(r) = \frac{\exp(-\Sigma r)}{4\pi r^2} S_p. \tag{2.9}$$

Nuclide Densities

Both factors in Eq. (2.5)—the nuclide density N and the microscopic cross section σ—require further discussion. First, consider the densities. Avogadro's number, $N_0 = 0.6023 \cdot 10^{24}$, is the number of molecules in one gram molecular weight of a substance, Thus if A is the molecular weight, N_0/A is the number of molecules in 1g of the substance. If we designate ρ as the density in grams/cm^3, then

$$N = \rho N_0/A \tag{2.10}$$

is the number of molecules/cm^3. Equation (2.5) becomes

$$\Sigma = \frac{\rho N_0}{A} \sigma, \tag{2.11}$$

where the density is in grams/cm^3 and σ is in cm^2. Usually the microscopic cross sections are tabulated in barns (designated as b) where $1 \mathrm{b} = 10^{-24}$ cm^2.

In many cases the formulas above may be applied directly to a chemical element even though mixtures of isotopes are included, provided the cross sections are measured for the elements as they exist in nature. Thus, for example, we treat iron as a single cross section without specifying the isotope, even though it has a molecular weight of 55.8 because the isotopes iron—54, 56, and 57—all are present in significant amounts. In situations where the cross sections are measured from particular isotopes, A in the foregoing equations is the atomic weight of the particular isotope.

In reactor physics the need sometimes arises to express the cross section of an element in terms of cross sections of its constituent isotopes. To accomplish this we first let N^i/N denote the atomic fraction of the isotope with atomic weight A_i. The atomic weight of the mixture is then

$$A = \sum_i (N^i/N)A_i, \tag{2.12}$$

where $N = \sum_i N^i$ and the macroscopic cross section of the combination of isotopes may be written as

$$\Sigma = \frac{\rho N_o}{A} \sum_i \frac{N^i}{N} \sigma^i, \tag{2.13}$$

where σ^i is the microscopic cross section of the ith isotope.

To compute the cross sections of molecules, the cross sections of the number of atoms of each element in the molecule must be included. Thus for water, with molecular weight 18, account must be taken for the number of hydrogen and oxygen atoms:

$$\Sigma^{H_2O} = \frac{\rho_{H_2O} N_o}{18} \left(2\sigma^H + \sigma^O \right). \tag{2.14}$$

We may define a composite microscopic cross section for a molecule, in the case of water

$$\sigma^{H_2O} = 2\sigma^H + \sigma^O, \tag{2.15}$$

so that Eq. (2.14) simplifies to $\Sigma^{H_2O} = N_{H_2O} \sigma^{H_2O}$ with $N_{H_2O} = \rho_{H_2O} N_o / 18$.

Frequently, materials are combined by volume fractions. Let V_i be the volumes, and V_i / V the volume fractions, where $V = \sum_i V_i$. The cross section for the mixture is then

$$\Sigma = \sum_i (V_i / V) N_i \sigma^i, \tag{2.16}$$

where each of the nuclide number densities is given by

$$N_i = \rho_i N_0 / A_i, \tag{2.17}$$

and ρ_i and A_i are the densities and atomic weights corresponding to a nuclide with a microscopic cross section of σ^i. Equation (2.16) may, of course, also be written in terms of the macroscopic cross sections of the constituents:

$$\Sigma = \sum_i (V_i / V) \Sigma^i, \tag{2.18}$$

where $\Sigma^i = N_i \sigma^i$. Sometimes mixtures are given in terms of mass fractions. We treat such situations by combining Eqs. (2.16) and (2.17) to write:

$$\Sigma = \sum_i (M_i / M) \frac{\rho N_o}{A_i} \sigma^i, \tag{2.19}$$

where $M_i / M = \rho_i V_i / \rho V$ is the mass fraction, $M = \sum_i M_i$, and the density is given by $\rho = M / V$.

Enriched Uranium

The cross sections designated for uranium are for natural uranium, which consists of 0.7% uranium-235 and 99.3% uranium-238. Frequently, however, designers call for enriched uranium in order to increase the ratio of fissile to fertile material. Enrichment may be defined in two ways. Atomic enrichment is the ratio of uranium-235 atoms to the total number of uranium atoms. Using the shorthand notation for fissile and fertile isotopes introduced in Section 1.6, the atomic enrichment is

$$\tilde{e}_a = N^{25}/(N^{25} + N^{28}),\tag{2.20}$$

and hence $1 - \tilde{e}_a = N^{28}/(N^{25} + N^{28})$. Inserting these expressions into Eqs. (2.12) and (2.13) yields a uranium cross section of

$$\Sigma^U = \frac{\rho_U N_o}{\tilde{e}_a 235 + (1 - \tilde{e}_a)238}\left[\tilde{e}_a\sigma^{25} + (1 - \tilde{e}_a)\sigma^{28}\right].\tag{2.21}$$

Alternately, mass (or weight) enrichment is the ratio of the mass of uranium-235 to the total uranium mass:

$$\tilde{e}_w = M^{25}/(M^{25} + M^{28}),\tag{2.22}$$

and correspondingly $1 - \tilde{e}_w = M^{28}/(M^{25} + M^{28})$. Then from Eq. (2.19) the uranium cross section is

$$\Sigma^U = \rho_U N_o\left[\frac{1}{235}\tilde{e}_w\sigma^{25} + \frac{1}{238}(1 - \tilde{e}_w)\sigma^{28}\right].\tag{2.23}$$

The two enrichments are often quoted as atom percent (a/o) and weight percent (w/o), respectively. They are closely related. Noting that $N_i = \rho_i N_o/A_i$ and $M_i = \rho_i V$, we may eliminate the densities between Eqs. (2.20) and (2.22) to obtain:

$$\tilde{e}_a = (1 + 0.0128\tilde{e}_w)^{-1}1.0128\tilde{e}_w.\tag{2.24}$$

Thus if we take $\tilde{e}_w = 0.00700$ for natural uranium, then $\tilde{e}_a = 0.00709$, and the fractional differences become smaller for higher enrichments. Except were very precise calculations are called for, we may ignore these small differences and allow both Eqs. (2.21) and (2.23) to be simplified to

$$\Sigma^U \simeq \frac{\rho_U N_o}{238}\sigma^U\tag{2.25}$$

with uranium's microscopic cross section approximated by

$$\sigma^U = \tilde{e}\sigma^{25} + (1 - \tilde{e})\sigma^{28}.\tag{2.26}$$

Except were stated otherwise, the pages that follow will take $\tilde{e}$ to be the atom enrichment, and utilize Eqs. (2.12) and (2.13) for determining uranium cross sections.

Cross Section Calculation Example

Frequently more than one of the foregoing formulas must be used in combination to obtain a macroscopic cross section. For example, suppose we want to calculate the cross sections for 8% enriched uranium dioxide (UO_2) that is mixed in a 1:3 volume ratio with graphite (C). The basic data required are the microscopic cross sections of the uranium isotopes, and of oxygen and carbon: $\sigma^{25} = 607.5\,b$, $\sigma^{28} = 11.8\,b$, $\sigma^{O} = 3.5\,b$, $\sigma^{C} = 4.9\,b$. We also need the densities of UO_2 and carbon: $\rho_{UO_2} = 11.0\,g/cm^3$, $\rho_C = 1.60\,g/cm^3$.

We first calculate the composite microscopic cross section for 8% enriched uranium. From Eq. (2.26) we have

$$\sigma^{U} = 0.08 \cdot 607.5 + (1 - 0.08) \cdot 11.8 = 59.5\,b.$$

The microscopic cross section of UO_2 is then

$$\sigma^{UO_2} = 59.5 + 2 \cdot 3.5 = 66.5\,b.$$

Noting that $1\,b = 10^{-24}\,cm^2$, the macroscopic cross section of the enriched UO_2 is

$$\Sigma^{UO_2} = \frac{11 \cdot 0.6023 \cdot 10^{24}}{238 + 2 \cdot 16} 66.5 \cdot 10^{-24} = 1.63\,cm^{-1}.$$

The macroscopic cross section of carbon is

$$\Sigma^{C} = \frac{1.6 \cdot 0.6023 \cdot 10^{24}}{12} 4.9 \cdot 10^{-24} = 0.39\,cm^{-1}.$$

Since UO_2 and C are mixed in a 1:3 ratio by volume, form Eq. (2.18) we obtain

$$\Sigma = \frac{1}{4}\Sigma^{UO_2} + \frac{3}{4}\Sigma^{C} = \frac{1}{4}1.63 + \frac{3}{4}0.39 = 0.70\,cm^{-1}.$$

Reaction Types

Thus far we have considered only the probability that a neutron has made a collision, without consideration of what happens subsequently. The cross section that we have been dealing with is designated as the total cross section, and often denoted with a subscript t: σ_t. Upon striking a nucleus, the neutron is either scattered or it is absorbed. The relative likelihoods of a scattering or an absorption are represented by dividing the total cross section into scattering and absorption cross sections:

$$\sigma_t = \sigma_s + \sigma_a. \tag{2.27}$$

Given a collision, σ_s/σ_t is the probability that the neutron will be scattered and σ_a/σ_t the probability that it will be absorbed. Scattering may be either elastic or inelastic. Thus in the most general case we may divide the scattering cross section to read

$$\sigma_s = \sigma_n + \sigma_{n'}. \tag{2.28}$$

Here σ_n denotes the elastic scattering cross section. Elastic scattering conserves both momentum and kinetic energy; it may be modeled as a billiard ball collision between a neutron and a nucleus. In an inelastic scattering collision, with cross section denoted by $\sigma_{n'}$, the neutron gives up some of its energy to the nucleus, leaving it in an excited state. Thus while momentum is conserved in an inelastic collision, kinetic energy is not; the nucleus gives up excitation energy by emitting one or more gamma rays along with the neutron.

In its simplest form, the absorption reaction creates a compound nucleus in an excited state. But instead of reemitting a neutron it eliminates the excitation energy by emitting one or more gamma rays. This is referred to as a capture reaction, and denoted by σ_γ. In many cases the new isotope thus created is not stable and will later undergo radioactive decay. In a fissionable material, following neutron absorption, the neutron may simply be captured, or it may cause fission. For fissionable materials we thus divide the absorption cross section as

$$\sigma_a = \sigma_\gamma + \sigma_f, \tag{2.29}$$

where σ_f is the fission cross section. We again may make a probabilistic interpretation: Given a neutron absorption, σ_γ/σ_a is the probability that the neutron will be captured and σ_f/σ_a the probability that a fission will result.

We express macroscopic cross sections for particular reaction types by using Eq. (2.5) in the same way as before. Suppose we let $x = s, a, \gamma, f$ signify scattering, absorption, capture, fission, and so on. Then we may write

$$\Sigma_x = N\sigma_x, \tag{2.30}$$

and analogous modifications may be made by adding these subscripts to preceding equations for microscopic and macroscopic cross sections.

From the foregoing equations we may also easily show that macroscopic cross sections for different reaction types add in the same way as microscopic cross sections. Thus analogous to Eq. (2.27) we have $\Sigma_t = \Sigma_s + \Sigma_a$ and so on.

2.3   Neutron Energy Range

Thus far we have not discussed the dependence of cross sections on neutron kinetic energy. To take energy into account we write each of the above cross sections as functions of energy by letting $\sigma_x \rightarrow \sigma_x(E)$ and similarly, as a result of Eq. (2.30), $\Sigma_x \rightarrow \Sigma_x(E)$. The energy dependence of cross sections is fundamental to neutron behavior in chain reactions and thus warrants detailed consideration. We begin by establishing the upper and lower limits of neutron energies found in fission reactors.

Neutrons born in fission are distributed over a spectrum of energy. Defining $\chi(E)dE$ as the fraction of fission neutrons born with energies between E and $E + dE$, a reasonable approximation to the fission spectrum is given by

$$\chi(E) = 0.453 \exp(-1.036E) \sinh\left(\sqrt{2.29E}\right), \qquad (2.31)$$

where E is in MeV and $\chi(E)$ is normalized to one:

$$\int_0^\infty \chi(E)dE = 1. \qquad (2.32)$$

The logarithmic energy plot of Fig. 2.2 shows the fission spectrum, $\chi(E)$. Fission neutrons are born in the MeV energy range with

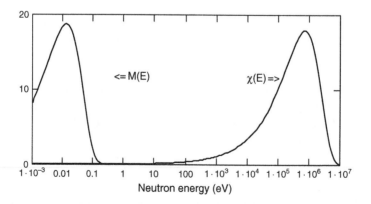

FIGURE 2.2   Fission and thermal neutron energy spectra.

an average energy of about 2 MeV, and the most probable energy is about 3/4 MeV. The numbers of fission neutrons produced with energies greater than 10 MeV is negligible, which sets the upper limit to the energy range of neutrons in reactors.

Neutrons born in fission typically undergo a number of scattering collisions before being absorbed. A neutron scattering from a stationary nucleus will transfer a part of its momentum to that nucleus, thus losing energy. However at any temperature above absolute zero, the scattering nuclei will possess random thermal motions. According to kinetic theory, the mean kinetic energy of such nuclei is

$$\bar{E} = \frac{3}{2}kT, \tag{2.33}$$

where k is the Boltzmann constant and T is the absolute temperature. For room temperature of $T = 293.61$ K the mean energy amounts to 0.0379 eV. Frequently, thermal neutron measurements are recorded at 1.0 kT, which, at room temperature, amounts to 0.0253 eV. In either case these energies are insignificant compared to the MeV energies of fission neutrons. Thus the scatting of neutrons causes them to lose kinetic energy as they collide with nearly stationary nuclei until they are absorbed or are slowed down to the eV range. In the idealized situation where no absorption is present, the neutrons would eventually come to equilibrium with the thermal motions of the surrounding nuclei. The neutrons would then take the form of the famed Maxwell-Boltzmann distribution

$$M(E) = \frac{2\pi}{(\pi kT)^{3/2}} E^{1/2} \exp(-E/kT), \tag{2.34}$$

where E is in eV, Boltzmann's constant is $k = 8.617065 \times 10^{-5}$ eV/K, and $M(E)$ is normalized to one:

$$\int_0^\infty M(E)dE = 1. \tag{2.35}$$

Figure 2.2 shows $M(E)$ along with $\chi(E)$ to indicate the energy range over which neutrons may exist in a nuclear reactor. Realize, however, that some absorption will always be present. As a result the spectrum will be shifted upward somewhat from $M(E)$ since the absorption precludes thermal equilibrium from ever being completely established. The fraction of neutrons with energies less than 0.001eV in the room temperature Maxwell-Boltzmann distribution is quite small, and we thus take it as the lower bound of energies that we need to consider. In general, we may say the primary range of interest for neutron in a chain reactor is in the range 0.001 eV $< E <$ 10 MeV. Thus the neutron energies range over roughly 10 orders of magnitude!

For descriptions of neutron cross sections important to reactor physics, it is helpful to divide them into to three energy ranges. We refer to fast neutrons as being those with energies over the range where significant numbers of fission neutrons are emitted: 0.1 MeV $< E <$ 10 MeV. We call thermal neutrons those with small enough energies that the thermal motions of the surrounding atoms can significantly affect their scattering properties: 0.001 eV $< E <$ 1.0 eV. We lump all the neutrons in between as epithermal or intermediate energy neutrons: 1.0 eV $< E <$ 0.1 MeV.

2.4 Cross Section Energy Dependence

We begin our description of the energy dependence of cross sections with hydrogen; since it consists of a single proton, its cross section is easiest to describe. Hydrogen has only elastic scattering and absorption cross sections. Since it has no internal structure, hydrogen is incapable of scattering neutrons inelastically. Figure 2.3a is a plot of hydrogen's elastic scattering cross section. The capture cross section, shown in Fig. 2.3b, is inversely proportional to $\sqrt{E}$, and since energy is proportional to the square of the speed, it is referred to as a $1/v$ or "one-over-v" cross section. Hydrogen's capture cross section—which is the same as absorption since there is no fission—is only large enough to be of importance in the thermal energy range. The absorption cross section is written as

$$\sigma_a(E) = \sqrt{E_o/E}\,\sigma_a(E_o). \tag{2.36}$$

Conventionally, the energy is evaluated at $E_o = kT$, in combination with the standard room temperature of $T = 293.61$ K. Thus

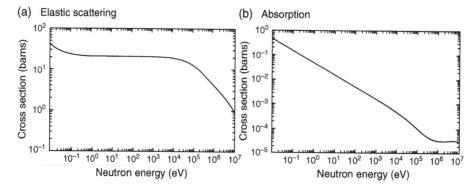

FIGURE 2.3 Microscopic cross sections of hydrogen-1 (from http://www. dne.bnl.gov/CoN/index.html). (a) Elastic scattering, (b) Absorption.

$E_o = 0.0253$ eV. For most purposes we may ignore the low- and high-energy tails in the scattering cross section. The total cross section may then be approximated as

$$\sigma_t(E) = \sigma_s + \sqrt{E_o/E}\,\sigma_a(E_o). \tag{2.37}$$

Hydrogen-2, or deuterium, cross sections have an analogous behavior, except that the scattering cross section is moderately larger, and the absorption cross section much smaller.

Like hydrogen, other nuclei have elastic scattering cross sections, which may be equated to simple billiard ball collisions in which kinetic energy is conserved. These are referred to as potential scattering cross sections because the neutron scatters from the surface of the nucleus, rather than entering its interior to form a compound nucleus. Potential scattering cross sections are energy independent except at very low or high energies. Their magnitude is directly proportional to the cross-sectional area of the nucleus, where the radius of the nucleus may be given in terms of the atomic weight as $R = 1.25 \times 10^{-13} A^{1/3}$ cm. Further understanding of neutron cross sections, however, requires that we examine reactions resulting from the formation of compound nuclei.

Compound Nucleus Formation

If a neutron enters a nucleus—instead of scattering from its surface as in potential scattering—a compound nucleus is formed, and it is in an excited state. There are two contributions to this excitation energy. The first derives from the kinetic energy of the neutron. We determine excitation energy as follows. Suppose a neutron of mass m and velocity v hits a stationary nucleus of atomic weight A and forms a compound nucleus. Conservation of momentum requires that

$$mv = (m + Am)\mathrm{V}. \tag{2.38}$$

Kinetic energy, however, is not conserved the formation. The amount lost is

$$\Delta E_{ke} = \frac{1}{2}mv^2 - \frac{1}{2}(m + Am)\mathrm{V}^2, \tag{2.39}$$

where V is the speed of the resulting compound nucleus. Eliminating V between these equations then yields

$$\Delta E_{ke} = \frac{A}{1+A}\frac{1}{2}mv^2, \tag{2.40}$$

which may be shown to be identical to the neutron kinetic energy before the collision measured in the center of mass system. Hence we hereafter denote it by E_{cm}. The second contribution to the excitation energy is the binding energy of the neutron, designated by E_B. The excitation energy of the compound nucleus is $E_{cm} + E_B$. Note that even very slow moving thermal neutrons will excite a nucleus, for even though $E_{cm} \ll E_B$, the binding energy by itself may amount to a MeV or more.

The effects of the excitation energy on neutron cross sections relate strongly to the internal structure of the nucleus. Although the analogy is far from complete, these effects can be roughly understood by comparing atomic to nuclear structures. The electrons surrounding a nucleus are in distinct quantum energy states and can be excited to higher states by imparting energy from the outside. Likewise, the configurations of nucleons that form a nucleus are in quantum states, and the addition of a neutron accompanied by its kinetic energy create a compound nucleus that is in an excited state. Following formation of a compound nucleus one of two things happen: the neutron may be reemitted, returning the target nucleus to its ground state; this scattering is elastic, even though a compound nucleus was formed temporarily in the process. Alternately, the compound nucleus may return to its ground state by emitting one or more gamma rays; this is a neutron capture reaction through which the target nucleus is transmuted to a new isotope as the result of the neutron gained.

With higher incoming neutron energies the compound nucleus may gain sufficient excitation energy to emit both a lower energy neutron and a gamma ray; thus inelastic scattering results, and at yet higher energies other reactions may result as well. In fissile and fertile materials, of course, the fission reaction is the most important consequence of compound nucleus formation. Before considering these reactions in detail, we first examine the resonance structure of compound nuclei and the effect that it has on scattering and absorption cross sections.

Resonance Cross Sections

The likelihood of compound nucleus formation greatly increases if the excitation energy brought by the incident neutron corresponds to a quantum state of the resulting nuclei. Scattering and absorption cross sections exhibit resonance peaks at neutron kinetic energies corresponding to those quantum states. Figure 2.4 illustrates the peaks in the scattering and capture cross sections of sodium-23. Each nuclide has its own unique resonance structure, but generally the heavier a nucleus is, the more energy states it will have, and they will be more closely packed together. Figure 2.5 illustrates this progression of state packing using carbon, aluminum, and uranium isotopes as examples. The correlation between quantum state density and atomic weight results in the resonance of lighter nuclides beginning to occur only at higher energies. For example, the lowest resonance in carbon-12 occurs at 2 MeV, in oxygen-16 at 400 keV, in sodium-23 at 3 keV, and in uranium-238 at 6.6 eV. Likewise the resonances of lighter nuclei are more widely spaced and tend to

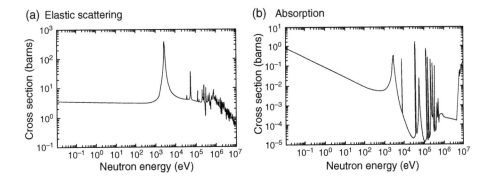

FIGURE 2.4   Microscopic cross sections of sodium-23 (from http://www.dne.bnl.gov/CoN/index.html). (a) Elastic scattering, (b) Absorption.

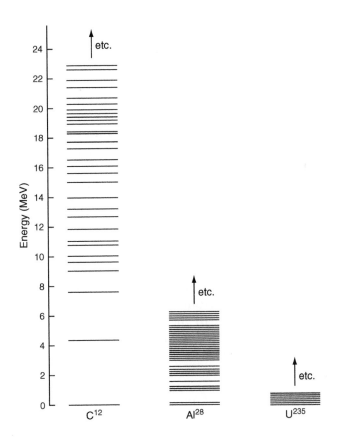

FIGURE 2.5   Energy levels of carbon-12, aluminum-28, and uranium-235. Adapted from *Introduction to Nuclear Reactor Theory*, 1972, by John R. Lamarsh. Copyright by the American Nuclear Society, La Grange Park, IL.

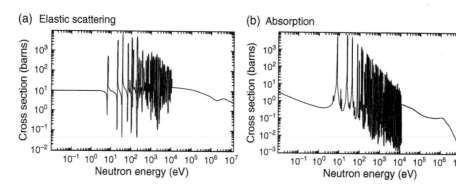

FIGURE 2.6 Microscopic cross sections of uranium-238 (from http://www.dne.bnl.gov/CoN/index.html). (a) Elastic scattering, (b) Absorption.

have a smaller ratio of capture to scattering cross section. Comparing the cross sections of uranium-238 given in Fig. 2.6 with those of sodium-23 in Fig. 2.4 illustrates these trends in resonance structure.

A noteworthy feature of the uranium cross sections in Fig. 2.6 is that the resonances appear to suddenly stop at approximately 10 keV. In fact, they extend to higher energies but are so tightly packed that at present they cannot be resolved experimentally. Thus the apparently smooth curve conceals unresolved resonances at higher energies. These must be treated by statistical theory until more refined experiments are able to resolve them. The situation is similar for other heavy nuclides.

Nuclear theory predicts that the energy dependence of the cross sections in the vicinity of each resonance will take the form of the Breit-Wigner formula. For the capture cross sections

$$\sigma_\gamma(E) = \sigma_o \frac{\Gamma_\gamma}{\Gamma} \left(\frac{E_r}{E}\right)^{1/2} \frac{1}{1 + 4(E - E_r)^2/\Gamma^2}, \qquad (2.41)$$

where E_r is the resonance energy, and Γ is approximately equal to the width of the resonance at half of the cross section's maximum value. In general, not all of the neutrons that collide near a resonance energy will be captured; some will be reemitted in resonance elastic scattering. The elastic scattering cross section in the vicinity of the resonance has three contributions:

$$\sigma_n(E) = \sigma_o \frac{\Gamma_n}{\Gamma} \frac{1}{1 + 4(E - E_r)^2/\Gamma^2} + \sigma_o \frac{2R}{\lambda_o} \frac{2(E - E_r)/\Gamma}{1 + 4(E - E_r)^2/\Gamma^2} + 4\pi R^2.$$

$$(2.42)$$

The first term is the resonance scattering, while the third is the energy-independent potential scattering. The second term arises from a quantum mechanical interference effect between resonance and potential scattering. In heavier nuclei, such as uranium, the interference is visible as a dip in the scattering cross section just below the resonance energy. For nonfissionable materials $\Gamma = \Gamma_\gamma + \Gamma_n$, and thus σ_0 is the resonance cross section when $E = E_r$, and λ_0 is the reduced neutron wavelength. In reactor problems distinguishing between resonance and potential scattering is sometimes advantageous. We do this by writing Eq. (2.42) as

$$\sigma_n(E) = \sigma_{nr}(E) + \sigma_{np}, \qquad (2.43)$$

where the first two terms of Eq. (2.41) are included in the resonance contribution, σ_{nr}, and the third constitutes the potential scattering, σ_{np}.

No discussion of resonance cross sections is complete without a description of Doppler broadening. Strictly speaking, neutron cross sections are written in terms of the relative speed between neutron and nucleus in the center of mass system. Normally, the kinetic energy of the incident neutron is so much larger than that of the nucleus, which is caused only by thermal motion, that the nucleus can be assumed to be stationary. Thus the cross section formulas above do not account for the thermal motions of the target nuclei. If the cross sections are a relatively smooth function of energy, these motions are unimportant. However, when cross sections are sharply peaked, as they are for the resonances described by the Breit-Wigner formulas, the formulas must be averaged over the range of relative speeds characterized as a function of temperature by the Maxwell-Boltzmann distribution of atom velocities. This averaging has the net effect of slightly smearing the resonances in energy, making them appear wider and less peaked. The smearing becomes more pronounced with increased temperature, as shown in exaggerated form for the resonance capture cross section curve of Fig. 2.7. The importance of Doppler

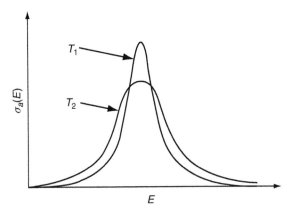

FIGURE 2.7   Doppler broadening of resonance capture cross section, with $T_1 < T_2$.

broadening in providing negative temperature feedback and thus stability to nuclear reactors is discussed in Chapter 9.

Threshold Cross Sections

With higher incident neutron energies—and as a result higher excitation energies—additional reactions become possible. These we refer to as threshold reactions because the cross section is zero below the threshold energy. Inelastic scattering cross sections exhibit threshold behavior because for such scattering to occur the incident neutron must have enough kinetic energy both to raise the target nucleus to an excited quantum state and to overcome the binding energy and be reemitted. Referring again to the examples of Fig. 2.5 we note that the lowest excited state of a nucleus generally decreases with increasing atomic weight. As a result the threshold for inelastic scattering also decreases with increasing atomic number. For the lighter nuclides, inelastic scattering thresholds are so high that the reaction is insignificant in reactors: The threshold for carbon-12 is 4.8 MeV, whereas that for oxygen-16 is 6.4 MeV. However, for heavier elements the threshold is lower; in uranium-238 is it at 0.04 MeV. Fertile materials, such as uranium-238, also have thresholds above which fission becomes possible; the threshold for uranium-238 fission is approximately 1.0 MeV. Figure 2.8 depicts the threshold cross section for both the inelastic scattering and fission in uranium-238.

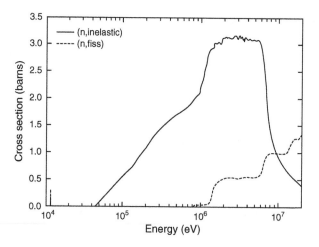

FIGURE 2.8 Microscopic threshold cross sections for uranium-238 (courtesy of W. S. Yang, Argonne National Laboratory).

A third class of threshold reaction that emits neutrons is $(n, 2n)$ in which the incident neutron ejects two neutrons from a nuclide. However, the threshold for this reaction is sufficiently high and the cross section small enough that generally it can be ignored in elementary treatments of reactor physics.

Fissionable Materials

Fissionable isotopes are either fissile or fertile, as discussed in Chapter 1. Incident neutrons of any energy cause fission in a fissile material. Figure 2.9 depicts the fission cross section of urainium-235, the only fissile material that occurs naturally. Uranium-238, which makes up 99.3% of natural uranium, fissions only from neutrons with an incident energy of a MeV or more as illustrated by the threshold in its fission cross section plotted in Fig. 2.8. It is, however, fertile, for following neutron capture it decays according to Eq. (1.28) to plutonium-239, which is fissile. Figure 2.10 shows the fission cross section of plutonium-239. If plutonium-239 captures an additional neutron instead of fissioning, it becomes plutonium-240, which is also a fertile isotope, for and if it captures an additional neutron it becomes plutonium-241, which is fissile. In addition to uranium-238, thorium-232 is a naturally occurring fertile isotope, for following neutron capture it undergoes radioactive decay to become uranium-233, which is fissile. The fission cross section of uranium-233 appears similar to those plotted in Figs. 2.9 and 2.10.

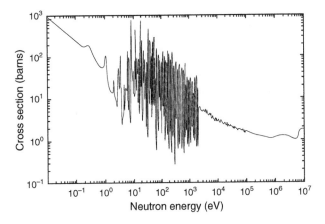

FIGURE 2.9 Microscopic fission cross sections of uranium-235 (from http://www.dne.bnl.gov/CoN/index.html).

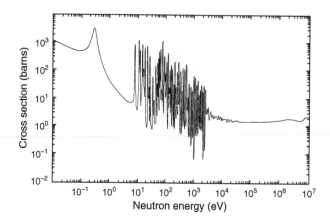

FIGURE 2.10 Microscopic fission cross sections of plutonium-239 (from http://www.dne.bnl.gov/CoN/index.html).

2.5 Neutron Scattering

The neutron energy spectrum in a reactor lies between the extremes of fission and thermal equilibrium. It is determined largely by the competition between scattering and absorption reactions. For neutrons with energies significantly above the thermal range, a scattering collision results in degradation of the neutron energy, whereas neutrons near thermal equilibrium may either gain or lose energy in interacting with the thermal motions of the nuclei of the surrounding media. Energy degradation caused by scattering is referred to as neutron slowing down. In a medium for which the average energy loss per collision and the ratio of scattering to absorption cross section is large, the neutron spectrum will be close to thermal equilibrium and is then referred to as a soft or thermal spectrum. Conversely, in a system with small ratios of neutron degradation to absorption, neutrons are absorbed before significant slowing down takes place. The neutron spectrum then lies closer to the fission spectrum and is said to be hard or fast. To gain a more quantitative understanding of neutron energy distributions we consider first elastic and then inelastic scattering. Recall that in elastic scattering mechanical energy is conserved, that is, the sums of the kinetic energies of the neutron and the target nucleus are the same before and after the collision. In inelastic scattering the neutron leaves the target nucleus in an excited—that is, more energetic—state. Thus the sum of the neutron and nucleus kinetic energies following the collision is less than before by the amount of energy deposited to form the excited state. Both elastic and inelastic scattering are of considerable importance in nuclear reactors. We treat elastic scattering first.

Elastic Scattering

For simplicity we first consider the head-on collision between a neutron with speed v and a stationary nucleus of atomic mass A. If we take m as the neutron mass then the nuclear mass will be approximately Am. If v' and V are the neutron and nucleus speeds after the collision then conservation of momentum yields

$$m \cdot v = m \cdot v' + (Am)V, \qquad (2.44)$$

while from conservation of mechanical energy

$$\frac{1}{2}m \cdot v^2 = \frac{1}{2}m \cdot v'^2 + \frac{1}{2}(Am)V^2. \qquad (2.45)$$

Letting E and E' be the neutron energy before and after the collision, we may solve these equations to show that the ratio of neutron energies is

$$\frac{E'}{E} = \left(\frac{A-1}{A+1}\right)^2. \qquad (2.46)$$

Clearly the largest neutron energy losses result from collisions with light nuclei. A neutron may lose all of its energy in a collision with a hydrogen nucleus, but at the other extreme, it can lose no more than 2% of its energy as the result of an elastic collision with uranium-238.

Of course, head-on collisions cause the maximum neutron energy loss, although in reality most neutrons will make glancing collisions in which they are deflected and lose a smaller part of their energy. If elastic scattering is analyzed not in the laboratory but in the center of mass system as a collision between two spheres, all deflection angles are equally likely, and the scattering is said to be isotropic in the center of mass system (or often just isotropic). Detailed analyses found in more advanced texts result in a probability distribution for neutron energies following a collision. Suppose a neutron scatters elastically at energy E. Then the probability that its energy following collision will be between E' and $E' + dE'$ will be

$$p(E \to E')dE' = \begin{cases} \dfrac{1}{(1-\alpha)E}dE', & \alpha E \leq E' \leq E, \\ 0 & \text{otherwise} \end{cases} \qquad (2.47)$$

where

$$\alpha = (A-1)^2 \big/ (A+1)^2. \qquad (2.48)$$

Often the need arises to combine the probability distribution for scattered neutrons with the scattering cross section. We then define

$$\sigma_s(E \rightarrow E') = \sigma_s(E)p(E \rightarrow E'), \tag{2.49}$$

where the corresponding macroscopic form is

$$\Sigma_s(E \rightarrow E') = \Sigma_s(E)p(E \rightarrow E'). \tag{2.50}$$

A similar expression applies to mixtures of nuclides:

$$\Sigma_s(E \rightarrow E') = \sum_i N_i\sigma_{si}(E \rightarrow E'), \tag{2.51}$$

where

$$\sigma_{si}(E \rightarrow E') = \sigma_{si}(E)p_i(E \rightarrow E'). \tag{2.52}$$

Alternately Eq. (2.50) is directly applicable, provided we define the composite scattering probability as

$$p(E \rightarrow E') = \frac{1}{\Sigma_s(E)}\sum_i \Sigma_{si}(E)p_i(E \rightarrow E'). \tag{2.53}$$

Slowing Down Decrement

The most widely employed measure of a nuclide's ability to slow neutrons down by elastic scattering is the slowing down decrement. It is defined as the mean value of the logarithm of the energy loss ratio or $\ln(E/E')$:

$$\xi \equiv \overline{\ln(E/E')} = \int \ln(E/E')p(E \rightarrow E')dE'. \tag{2.54}$$

Employing Eq. (2.47) once again we have

$$\xi = \int_{\alpha E}^{E} \ln(E/E')\frac{1}{(1-\alpha)E}dE', \tag{2.55}$$

which reduces to

$$\xi = 1 + \frac{\alpha}{1-\alpha}\ln\alpha. \tag{2.56}$$

The slowing down decrement is independent of the energy of the scattered neutron. Thus in elastic collisions the neutron loses on average the same logarithmic fraction of its energy, regardless of its initial energy, for ξ depends only on the atomic mass of the scattering nuclide. The slowing down decrement may be expressed in terms of the atomic mass of the scattering nuclide. Thus for $A = 1$, $\xi = 1$, and for $A > 1$ a reasonable approximation is

$$\xi \approx \frac{2}{A + 2/3}, \tag{2.57}$$

which gives an error of roughly 3% for $A = 2$, and successively smaller errors for larger values of A.

Using the definition of ξ we may make a rough estimate of the number n of elastic collisions required to slow a neutron from fission to thermal energy. Suppose we let $E_1, E_2, E_3, \ldots, E_n$ be the neutron energies after the first, second, third, and so on collisions. Then

$$\ln(E_0/E_n) = \ln(E_0/E_1) + \ln(E_1/E_2) + \ln(E_3/E_3) + \cdots + \ln(E_{n-1}/E_n). \tag{2.58}$$

Assuming that each of the n terms can be replaced by the average logarithmic energy loss ξ, we have

$$n = \frac{1}{\xi}\ln(E_0/E_n). \tag{2.59}$$

Taking fission energy as $E_0 = 2\,\text{MeV}$ and thermal energy as $E_n = 0.025\,\text{eV}$, we have $\ln(E_0/E_n) = \ln(2.0 \cdot 10^6/0.025) = 18.2$ and hence $n = 18.2/\xi$. Thus for hydrogen $n \approx 18$, for deuterium $(A = 2)$ $n \approx 25$, for carbon $(A = 12)$ $n \approx 115$, and for uranium-238 $n \approx 2275$. From this we observe that if we desire to slow neutrons down to thermal energies, light atomic weight materials are desirable components of a reactor core. Conversely, if fast neutrons are desired, lightweight materials should be avoided. Chapter 3 focuses considerable attention on elastic scattering, as well as the other properties of neutron cross sections, in examining their effects on the neutron energy spectra in reactors.

In situations where more than one nuclide is present, an average slowing down decrement may be derived by employing Eqs. (2.53) in Eq. (2.54):

$$\bar{\xi} = \frac{1}{\Sigma_s(E)} \sum_i \Sigma_{si}(E) \int \ln(E/E')p_i(E \to E')dE'. \tag{2.60}$$

Employing Eq. (2.47) for the scattering kernel with $\alpha \to \alpha_i$ then leads to

$$\bar{\xi} = \frac{1}{\Sigma_s} \sum_i \xi_i \Sigma_{si}, \tag{2.61}$$

where for brevity we have assumed energy-independent scattering cross sections.

Suppose, for example, we want to evaluate ξ_{H_2O}. In the denominator, we may use Eq. (2.14) for the scattering cross section of water directly, $\Sigma_s^{H_2O} = N_{H_2O}(2\sigma^H + \sigma^O)$. However, in the numerator we must use $\Sigma_s^H = 2N_{H_2O}\sigma_s^H$ and $\Sigma_s^O = N_{H_2O}\sigma^O$, the macroscopic cross sections of hydrogen and oxygen, separately. After cancellation of N_{H_2O} from numerator and denominator, Eq. (2.61) becomes

$$\xi_{H_2O} = \frac{2\xi_H\sigma_s^H + \xi_O\sigma_s^O}{2\sigma_s^H + \sigma_s^O}, \tag{2.62}$$

with $\sigma_s^H = 20\,\mathrm{b}$ and $\sigma_s^O = 3.8\,\mathrm{b}$, knowing that $\xi_H = 1$ and from Eq. (2.57) that $\xi_O = 2/(16 + 2/3) = 0.12$ we have $\xi_{H_2O} = (2 \cdot 1 \cdot \sigma_s^H + 0.12 \cdot \sigma_s^O)/(2 \cdot \sigma_s^H + \sigma_s^O) = 0.924$.

Inelastic Scattering

The situation for inelastic scattering is quite different. Elastic scattering cross sections are significant over the entire energy range of neutrons. But whereas low atomic weight nuclei cause large energy losses for elastic scattering, heavy isotopes do not, and so the effects of their elastic scattering on reactor physics are small. Conversely, as discussed earlier, only neutrons with energies above a threshold that is a characteristic of the target isotope can scatter inelastically. Moreover, these thresholds are low enough for significant inelastic scattering to occur only for the heavier atomic weight materials, such as uranium.

Inelastic scattering causes neutrons to lose substantial energy. The unique structure of energy levels that characterizes each nuclide, such as those illustrated in Fig. 2.5, determines the energies of inelastically scattered neutrons. To scatter inelastically the neutron must elevate the target nucleus to one of these states, from which it decays by emitting one or more gamma rays. The threshold for inelastic scattering is determined by the energy of the lowest excited state of the target nucleus, whereas the neutron's energy loss is determined predominantly by the energy level of the state that it excites. For

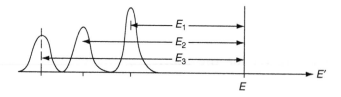

FIGURE 2.11   Inelastically scattered neutrons from energy E to E'.

example, if the neutron energy E is greater than the first three energy levels E_1, E_2, or E_3, then following the inelastic scatter the neutron would have energy $E' = E - E_1$, $E - E_2$, or $E - E_3$. This is illustrated in Fig. 2.11. The peaks, however, are slightly smeared over energy, since as in elastic scattering conservation of momentum requires that a neutron deflected through a larger angle will lose more energy than one deflected through a smaller angle. As the energy of the incident neutron increases, the spectrum of scattered neutrons can become quite complex if many states can be excited by the neutron's energy.

Bibliography

Bell, George I., and Samuel Glasstone, *Nuclear Reactor Theory*, Van Nostrand Reinhold, NY, 1970.

Cullen, D. E., "Nuclear Cross Section Preparation" *CRC Handbook of Nuclear Reactor Calculations*, *I*, Y. Ronen, ed., CRC Press, Boca Raton, FL, 1986.

Duderstadt, James J., and Louis J. Hamilton, *Nuclear Reactor Analysis*, Wiley, NY, 1976.

Jakeman, D., *Physics of Nuclear Reactors*, Elsevier, NY, 1966.

Mughabghab, S. F., *Atlas of Neutron Resonances*, Elsevier, Amsterdam, 2006.

Stacey, Weston M., *Nuclear Reactor Physics*, Wiley, NY, 2001.

Templin, L. J. ed., *Reactor Physics Constants*, 2nd ed., ANL-5800, Argonne National Laboratory, Argonne, IL, 1963.

Weinberg, A. M., and E. P. Wigner, *The Physical Theory of Neutron Chain Reactors*, University of Chicago Press, Chicago, 1958.

Problems

2.1. Neutrons impinge on a material with a cross section of $\Sigma = 0.8 \, \text{cm}^{-1}$. How thick must the material be if no more than 5.0% of the neutrons are to penetrate the material without making a collision? What fraction of the neutrons make their first collision within the first 2.0 cm of the material?

2.2. The uncollided flux at a distance r from a point source emitting neurons is given by Eq. (2.9).

    a. If you are 1 m away from a very small 1-Ci source of neutrons, what is the flux of neutrons in $n/\text{cm}^2/\text{s}$, neglecting scattering and absorption in air?

    b. If a shield is placed between you and the source, what absorption cross section would be required to reduce the flux by a factor of 10?

c. Suppose the shield made of the material specified in part b is only 0.5 m thick. How far must you be from the source, for the flux to be reduced by the same amount as in part b?

2.3. A material has a neutron cross section of $3.50 \times 10^{-24} \, \text{cm}^2/$ nuclei, and contains $4.20 \times 10^{23} \, \text{nuclei/cm}^3$.

a. What is the macroscopic cross section?
b. What is the mean free path?
c. If neutrons impinge perpendicularly on a slab of the material that is 3.0 cm thick what fraction of them will penetrate the slab without making a collision?
d. What fraction of the neutrons in part c will collide in the slab before penetrating a distance of 1.5 cm?

2.4. A boiling water reactor operates at 1000 psi. At that pressure the density of water and of steam are, respectively, $0.74 \, \text{g/cm}^3$ and $0.036 \, \text{g/cm}^3$. The microscopic cross sections of H and O thermal cross sections are 21.8 b and 3.8 b.

a. What is the macroscopic total cross section of the water?
b. What is the macroscopic total cross section of the steam?
c. If, on average, 40% of the volume is occupied by steam, then what is the macroscopic total cross section of the steam–water mixture?
d. What is the macroscopic total cross section of water under atmospheric conditions at room temperature?

2.5.* Determine the following:

a. The fraction of fission neutrons born with energies of less than 0.1 MeV.
b. The fraction of fission neutrons born with energies greater than 10 MeV.

2.6. Neutrons are distributed in the Maxwell-Boltzmann distribution given by Eq. (2.34):

a. Verify Eq. (2.35).
b. Verify Eq. (2.33).
c. Determine the most probable neutron energy.

2.7. How may parts per million of boron must be dissolved in water at room temperature to double its absorption cross section for thermal neutrons?

2.8. What is the total macroscopic thermal cross section of uranium dioxide (UO_2) that has been enriched to 4%? Assume $\sigma^{25} = 607.5 \, \text{b}$, $\sigma^{28} = 11.8 \, \text{b}$, $\sigma^O = 3.8 \text{b}$, and that UO_2 has a density of $10.5 \, \text{g/cm}^3$.

2.9. In the Breit-Wigner formula for the capture cross section show that Γ is equal to the width of the resonance at half height. What, if any, assumptions must you make to obtain this result?

2.10. Verify Eqs. (2.46) and (2.56).

2.11. Boron is frequently used as a material to shield against thermal neutrons. Using the data in Appendix E, estimate the thickness of boron required to reduce the intensity of a neutron beam by factors 100, 1000, 10,000, and 100,000.

2.12. A 5.0-cm-thick layer of purely absorbing material is found to absorb 99.90% of a neutron beam. The material is known to have a density of 4.0×10^{22} nuclei/cm^3. Determine the following:

   a. The macroscopic cross section.
   b. The mean free path.
   c. The microscopic cross section.
   d. Is the cross section as big as a barn?

2.13. Equal volumes of graphite and iron are mixed together. Fifteen percent of the volume of the resulting mixture is occupied by air pockets. Find the total macroscopic cross section given the following data: $\sigma_C = 4.75$ b, $\sigma_{Fe} = 10.9$ b, $\rho_C = 1.6$ g/cm^3, $\rho_{Fe} = 7.7$ g/cm^3. Is it reasonable to neglect the cross section of air? Why?

2.14. Neutrons scatter elastically at 1.0 MeV. After one scattering collision, determine the fraction of the neutrons that will have energies of less than 0.5 MeV if they scatter from the following:

   a. Hydrogen.
   b. Deuterium.
   c. Carbon-12.
   d. Uranium-238.

2.15. What is the *minimum* number of elastic scattering collisions required to slow a neutron down from 1.0 MeV to 1.0 eV in the following?

   a. Deuterium.
   b. Carbon-12.
   c. Iron-56.
   d. Uranium-238.

2.16. Using the macroscopic scattering cross sections in Appendix Table E-3, calculate the slowing down decrement for UO$_2$, where U is natural uranium. Does the presence of oxygen have a significant effect on the slowing down decrement?

2.17. Prove that $\xi = 1$ for hydrogen.

2.18. a. Show that for elastic scattering $\overline{E - E'} \equiv \int (E - E')p(E \rightarrow E')dE'$ is equal to $1/2(1 - \alpha)E$.
   b. Evaluate $\overline{E - E'}$ for ordinary water.

CHAPTER 3

Neutron Distributions in Energy

3.1 Introduction

In Chapter 1 we briefly introduced the concept of neutron multiplication, defining it as

$$k = \frac{\text{number of neutrons in } i\text{th} + 1 \text{ generation}}{\text{number of neutrons in } i\text{th generation}}, \tag{3.1}$$

where neutrons in a particular generation are considered to be born in fission, undergo a number of scattering collisions, and die in absorption collisions. Understanding what determines the magnitude of the multiplication is central to the study of neutron chain reactions. This chapter examines the determinants of multiplication with primary emphasis placed on the kinetic energy of the neutrons, for as we saw in Chapter 2, the basic data—the neutron cross sections—are strongly energy dependent. These energy dependencies define the two broad classes of reactors: thermal and fast. We first discuss the properties of nuclear fuel and of materials that moderate the neutron spectrum. With this background we proceed to provide a more detailed description of the energy distributions of neutrons in nuclear reactors, and then discuss the averaging of neutron cross sections over energy. We conclude by defining the neutron multiplication in terms of energy-averaged cross sections.

This chapter's discussions maintain two simplifications in order to focus the analysis on the energy variable. First, we assume that all neutrons are produced instantaneously at the time of fission, postponing discussion of the effects created by the small fraction of neutrons whose emission following fission is delayed until the detailed treatment of reactor kinetics in Chapter 5. Second, we defer analysis of the spatial distributions of neutrons in nuclear reactors to later chapters. For now we take the finite size of a reactor into account simply by noting that for many systems the multiplication can by approximated by

$$k = k_\infty P_{NL},\qquad(3.2)$$

where P_{NL} is the neutron nonleakage probability and k_∞ is the multiplication that would exist if a reactor's dimensions were infinitely large. Chapters 6 and 7 treat neutron leakage and other spatial effects in some detail. Our focus here is on neutron energy, and how the energy dependence of the cross sections dominates the determination of k_∞.

3.2   Nuclear Fuel Properties

Much of the physics of nuclear reactors is determined by the energy dependence of the cross sections of fissile and fertile materials over the range of incident neutron energies between the fission spectrum and the Maxwell-Boltzmann distribution of thermal neutrons, thus over the range between roughly 10 MeV and 0.001 eV. Recall that fissile nuclides have significant fission cross sections over this entire range as indicated by Figs. 2.9 and 2.10 for those of uranium-235 and plutonium-239. In contrast, fission of a fertile material is possible only for incident neutrons above some threshold; Fig. 2.8 indicates that the threshold for uranium-238 is at approximately 1.0 MeV. Not all of the neutrons absorbed by a fissile nucleus will cause fission. Some fraction of them will be captured, and that fraction is also energy dependent. As a result, the number of fission neutrons produced per neutron absorbed plays a central role in determining a reactor's neutron economy:

$$\eta(E) = \frac{\nu\Sigma_f(E)}{\Sigma_a(E)} = \frac{\text{fission neutrons produced}}{\text{neutrons absorbed}},\qquad(3.3)$$

where ν is the number of neutrons produced per fission, and

$$\Sigma_a(E) = \Sigma_\gamma(E) + \Sigma_f(E).\qquad(3.4)$$

To sustain a chain reaction, the average value of eta must be substantially more than one, for in a power reactor neutrons will be lost to absorption in structural, coolant, and other materials, and some will leak out of the system.

   To examine the behavior of $\eta(E)$ we first consider a single fissile isotope. We may then cancel the number densities from numerator and denominator of Eq. (3.3) to obtain

$$\eta(E) = \frac{\nu \sigma_f(E)}{\sigma_a(E)}. \tag{3.5}$$

From the plots of $\eta(E)$ shown in Fig. 3.1 for uranium-235 and pluto-nium-239 we see that concentrating neutrons at either high or low energies and avoiding the range between roughly 1.0 eV and 0.1 MeV where the curves dip to their lowest values most easily achieves a chain reaction. Except for naval propulsion systems designed for the military, however, fuels consisting predominately of fissile material are not employed in power reactors. Enrichment and fabrication costs would render them uneconomical. More importantly, the fuel would

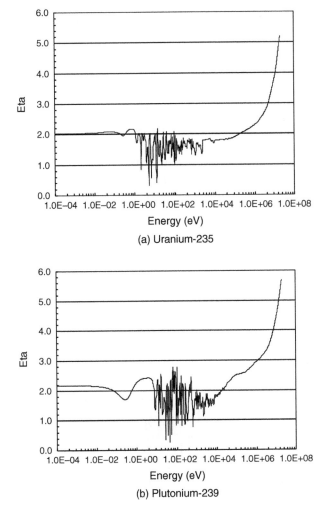

FIGURE 3.1 $\eta(E)$ for fissile isotopes (courtesy of W. S. Yang, Argonne National Laboratory). (a) Uranium-235, (b) Plutonium-249.

constitute weapons grade uranium or plutonium that would compound the problems of nuclear proliferation. Reactor fuels consist primarily of uranium-238 with a smaller fraction of fissile material, referred to as its enrichment. Depending on the design, civilian reactor fuels normally consist of uranium with enrichments ranging from the 0.7% of natural uranium up to approximately 20% fissile material.

To determine $\eta(E)$ for a reactor fuel, we first define the enrichment $\tilde{e}$ as the atom ratio of fissile to fissionable (i.e., fertile plus fissile) nuclei:

$$\tilde{e} = \frac{N_{fi}}{N_{fe} + N_{fi}}, \tag{3.6}$$

where *fi* and *fe* denote fissile and fertile. Equation (3.3) then reduces to

$$\eta(E) = \frac{\tilde{e}\nu\sigma_f^{fi}(E) + (1 - \tilde{e})\nu\sigma_f^{fe}(E)}{\tilde{e}\sigma_a^{fi}(E) + (1 - \tilde{e})\sigma_a^{fe}(E)}. \tag{3.7}$$

Figure 3.2 provides plots of $\eta(E)$ for natural (0.7%) and 20% enriched uranium. These curves illustrate the dramatic effect that the capture cross section of uranium-238 has in deepening the valley in $\eta(E)$ through the intermediate energy range. Conversely, above its threshold value 1.0 MeV, the increasing fission cross section of uranium-238 aids strongly in increasing the value of $\eta(E)$. The curves emphasize why power reactors are classified as fast or thermal

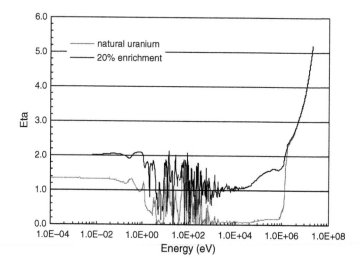

FIGURE 3.2 $\eta(E)$ for natural and 20% enriched uranium (courtesy of W. S. Yang, Argonne National Laboratory).

according to the energy spectra over which neutrons are concentrated, and why no intermediate spectrum reactors have been built.

Reactor designs must concentrate neutrons either in the fast or the thermal energy range and thus avoid the sharp valley that $\eta(E)$ exhibits over intermediate energies. As Chapter 2 emphasizes, scattering collisions cause neutrons to lose energy until they approach equilibrium in the thermal neutron range. Thus for fast reactor cores, designers eliminate materials other than the fuel as much as possible. They avoid low atomic weight materials in particular, for elastic scattering in such materials quickly reduces neutron energies to levels to where resonance capture in uranium-238 predominates. Even if all other materials could be eliminated, however, a fast reactor fueled with natural uranium is not possible, for the large inelastic scattering cross section of the 99.3% uranium-238 would then cause the fission neutrons to fall too quickly into the intermediate energy range. Consequently, fast reactors require fuels enriched to more than 10%.

For thermal reactors the situation is reversed. The reactor must contain a substantial quantity of low atomic weight material, referred to as a moderator. Its purpose is to slow down neutrons past the valley in $\eta(E)$ with relatively few collisions to the thermal energies where the fuel's ratio of neutron production to absorption again exceeds one by a substantial margin. With optimized ratios of moderator to fuel, thermal reactors can be designed with much lower enrichments than fast reactors; with some moderators—most notably graphite or heavy water—thermal reactors may be fueled with natural uranium. To understand why this is so we must examine the properties of moderators more closely.

3.3 Neutron Moderators

In thermal reactors moderator materials are required to reduce the neutron energies from the fission to the thermal range with as few collisions as possible, thus circumventing resonance capture of neutrons in uranium-238. To be an effective moderator a material must have a low atomic weight. Only then is ξ—the slowing down decrement defined by Eq. (2.54)—large enough to slow neutrons down to thermal energies with relatively few collisions. A good moderator, however, must possess additional properties. Its macroscopic scattering cross section must be sufficiently large. Otherwise, even though a neutron colliding with it would lose substantial energy, in the competition with other materials, too few moderator collisions would take place to have a significant impact on the neutron spectrum. Thus a second important parameter in determining a material's value as a moderator is the slowing down power, defined as $\xi \Sigma_s$,

TABLE 3.1
Slowing Down Properties of Common Moderators

Moderator	Slowing Down Decrement ξ	Slowing Down Power $\xi \Sigma_s$	Slowing Down Ratio $\xi \Sigma_s / \Sigma_a (thermal)$
H_2O	0.93	1.28	58
D_2O	0.51	0.18	21,000
C	0.158	0.056	200

where $\Sigma_s = N\sigma_s$ is the macroscopic scattering cross section. Note that the number density N must not be too small. Thus gases are eliminated. Helium, for example, has sufficiently large values of ξ and σ_s to be a good moderator but its number density is too small to have a significant impact on the energy distribution of neutrons in a reactor. Conversely, for the same reason gases such as helium may be considered as coolants for fast reactors since they do not degrade the neutron spectrum appreciably.

Table 3.1 lists values of the slowing down decrement and power for the three most common moderators. The table also includes the slowing down ratio: the ratio of the material's slowing down power to its thermal absorption cross section. If the thermal absorption cross section $\Sigma_a(E_{thermal})$ is large, a material cannot be used as a moderator; even though it may be effective in slowing down neutrons to thermal energy, it will then absorb too many of those same neutrons before they can make collisions with the fuel and cause fission. Note that heavy water has by far the largest slowing down ratio, followed by graphite and then by ordinary water. Power reactors fueled by natural uranium can be built using D_2O as the moderator. Because graphite has poorer moderating properties, the design of natural uranium fueled power reactors moderated by graphite is a more difficult undertaking. Reactors using a light water moderator and fueled with natural uranium are not possible; some enrichment of the uranium is required to compensate for the larger thermal absorption cross section of the H_2O.

Large thermal absorption cross sections eliminate other materials as possible moderators. For example, boron-10 has reasonable values of the slowing down decrement and power. Its thermal absorption cross section, however, is nearly 4000 b. As a result boron cannot be used as a moderator but is, in fact, one of the more common neutron "poisons," which are used to control or shut down the chain reactions.

The foregoing discussion focuses on elastic scattering, since inelastic scattering tends to be of much less importance in determining the energy distribution of neutrons in thermal reactors. The lighter weight materials either have no inelastic scattering cross section, or if they do, they are zero below a threshold that is quite high in energy. Fertile and fissile materials do have inelastic scattering above thresholds in the keV to MeV range. In thermal reactors inelastic collisions only modestly augment the slowing down by elastic scattering. The situation is quite different in fast reactors where the absence of moderator material causes inelastic scattering to become more important. Inelastic scattering of the fuel along with elastic scattering with the coolant and structural materials are the primary causes for unwanted energy spectrum degradation.

3.4 Neutron Energy Spectra

To recapitulate, the distribution of neutrons in energy is determined largely by the competition between scattering and absorption reactions. For neutrons with energies significantly above the thermal range, a scattering collision results in degradation of the neutron energy, whereas neutrons in thermal equilibrium have near equal probabilities of gaining or losing energy when interacting with the thermal motions of the nuclei that constitute the surrounding medium. In a medium for which the average energy loss per collision and the ratio of scattering to absorption cross section are both large, the neutron distribution in energy will be close to thermal equilibrium and is then referred to as a soft or thermal spectrum. Conversely in a system with small ratios of neutron degradation to absorption, neutrons are absorbed before significant slowing down takes place. The neutron distribution then lies closer to the fission spectrum and is said to be hard or fast.

The neutron distribution may be expressed in terms of the density distribution

$$\tilde{n}'''(E)dE = \begin{cases} \text{number of neutrons/cm}^3 \\ \text{with energies between } E \text{ and } E + dE, \end{cases} \tag{3.8}$$

which means that

$$n''' = \int_0^\infty \tilde{n}'''(E)dE = \text{total number of neutrons/cm}^3. \tag{3.9}$$

The more frequently used quantity, however, is the neutron flux distribution defined by

$$\varphi(E) = v(E)\tilde{n}'''(E), \tag{3.10}$$

where $v(E)$ is the neutron speed corresponding to kinetic energy E.

The flux, often called the scalar flux, has the following physical interpretation: $\varphi(E)dE$ is the total distance traveled during one second by all neutrons with energies between E and dE located in 1 cm^3. Likewise, we may interpret the macroscopic cross section as

$$\Sigma_x(E) = \left\{ \begin{array}{l} \text{Probability/cm of flight of a neutron} \\ \text{with energy } E \text{ undergoing a reaction of type } x. \end{array} \right. \quad (3.11)$$

Thus multiplying a cross section by the flux, we have

$$\Sigma_x(E)\varphi(E)dE = \left\{ \begin{array}{l} \text{Probable number of collisions of type } x/s/cm^3 \\ \text{for neutrons with energies between } E \text{ and } dE. \end{array} \right.$$
$$(3.12)$$

Finally, we integrate over all energy to obtain

$$\int_0^\infty \Sigma_x(E)\varphi(E)dE = \left\{ \begin{array}{l} \text{Probable number of collisions} \\ \text{of type } x/s/cm^3 \text{ of all neutrons.} \end{array} \right. \quad (3.13)$$

This integral is referred to as a reaction rate, or if $x = s$, a, f as the scattering, absorption, or fission rate.

A more quantitative understanding of neutron energy distributions results from writing down a balance equation in terms of the neutron flux. Since $\Sigma(E)\varphi(E)$ is the collision rate—or number of neutrons of energy E colliding/s/cm^3—each such collision removes a neutron from energy E either by absorption or by scattering to a different energy. We may thus regard it as a loss term that must be balanced by a gain of neutrons arriving at energy E. Such gains may come from fission and from scattering. The number coming from fission will be $\chi(E)$, given by Eq. (2.31). We next recall that the probability that a neutron that last scattered at energies between E' and $E' + dE'$ will be scattered to an energy E as $p(E' \rightarrow E)dE'$. Since the number of neutrons scattered from energy E' is $\Sigma_s(E')\varphi(E')$, the scattering contribution comes from integrating $p(E' \rightarrow E)\Sigma_s(E')\varphi(E')dE'$ over E'. The balance equation is thus

$$\Sigma_t(E)\varphi(E) = \int p(E' \rightarrow E)\Sigma_s(E')\varphi(E')dE' + \chi(E)s_f'''. \quad (3.14)$$

The specific form of $p(E' \rightarrow E)$ for elastic scattering by a single nuclide is given by Eq. (2.47), whereas Eq. (2.53) defines the composite probability for situations where the cross sections are sums over more than one nuclide. For brevity we write the foregoing equation as

$$\Sigma_t(E)\varphi(E) = \int \Sigma_s(E' \to E)\varphi(E')dE' + \chi(E)s_f''', \qquad (3.15)$$

where as in Eq. (2.50) we take $\Sigma_s(E' \to E) = p(E' \to E)\Sigma_s(E')$. The balance equation is normalized by the fission term, which indicates a rate of s_f''' fission neutrons produced/s/cm^3.

Using Eq. (3.15) to examine idealized situations over three different energy ranges provides some insight into the nature of neutron spectra, particularly of thermal reactors. First, we consider fast neutrons, whose energies are sufficient that $\chi(E)$ is significant. Normally the lower limit to this range is about 0.1 MeV. We then examine intermediate energy neutrons, which have energies below the range where fission neutrons are produced but sufficiently high that upscatter—that is, energy gained in a collision as a result of the thermal motion of the scattering nuclide—can be ignored. The lower cutoff for intermediate neutrons is conventionally taken as 1.0 eV. The intermediate energy range is often referred to as the resonance or slowing down region of the energy spectra because of the importance of these two phenomena. Third, we discuss slow or thermal neutrons defined as those with energies less than 1.0 eV; at the lower energies thermal motions of the surrounding nuclei play a predominant role in determining the form of the spectrum. In each of the three energy ranges general restrictions apply to Eq. (3.15). In the thermal and intermediate ranges no fission neutrons are born and thus $\chi(E) = 0$. In the intermediate and fast ranges there is no up-scatter, and therefore $\Sigma_s(E' \to E) = 0$ for $E' < E$.

Fast Neutrons

Over the energy range where fission neutrons are born both terms on the right of Eq. (3.15) contribute; near the top of that range the fission spectrum $\chi(E)$ dominates, since on average even one scattering collision will remove a neutron to a lower energy. In that case we may make the rough approximation,

$$\varphi(E) \approx \chi(E)s_f'''\big/\Sigma_t(E), \qquad (3.16)$$

which only includes the uncollided neutrons: those emitted from fission but yet to make a scattering collision. Even in the absence of moderators or other lower atomic weight materials the spectrum will be substantially degraded as a result of inelastic scattering collisions with uranium or other heavy elements. The presence of even small amounts of lighter weight materials, such as the metals used within a reactor core for structural support, adds to the degradation of the fast spectrum. Neutron moderators, of course greatly accelerate the slowing down of neutrons out of the fast range. In fast reactors, where lightweight materials are avoided, most of the neutrons are absorbed before scattering collisions slow them down below the low-energy tail of the fission spectrum.

Neutron Slowing Down

We next examine the energy range that extends below where $\chi(E)$ is significant but higher than the thermal energy range, where the thermal motions of the nuclei must be taken into account.

The Slowing Down Density

A useful concept for treating neutrons in this energy range is the slowing down density, which we define as

$$q(E) = \begin{cases} \text{number of neutrons slowing down} \\ \text{past energy } E/\text{s/cm}^3. \end{cases} \qquad (3.17)$$

At energies greater than where up-scatter occurs, any neutron produced by fission that is not absorbed at a higher energy must slow down past that energy. Thus

$$q(E) = -\int_E^\infty \Sigma_a(E')\varphi(E')dE' + \int_E^\infty \chi(E')dE'\,s_f''', \quad E > 1.0\,\text{eV}. \qquad (3.18)$$

In the intermediate range, below where fission neutron production is significant, the normalization of $\chi(E')$ given by Eq. (2.32) simplifies Eq. (3.18) to

$$q(E) = -\int_E^\infty \Sigma_a(E')\varphi(E')dE' + s_f''', \quad 1.0\,\text{eV} < E < 0.1\,\text{MeV}. \qquad (3.19)$$

Taking the derivative, we have

$$\frac{d}{dE}q(E) = \Sigma_a(E)\varphi(E). \qquad (3.20)$$

Thus $q(E)$ decreases as the neutrons slow down in proportion to the absorption cross section; if there is no absorption over some energy interval then the slowing down density remains constant.

In the intermediate range the primary form of absorption comes from the resonance capture cross sections discussed in Chapter 2. However, between those resonances the absorption cross section is small enough to be ignored. Thus between resonances we see from Eq. (3.20) that the slowing down density is independent of energy.

Moreover, since we are below the energies where fission neutrons are produced, with no absorption, Eq. (3.14) simplifies to

$$\Sigma_s(E)\varphi(E) = \int p(E' \rightarrow E)\Sigma_s(E')\varphi(E')dE'. \tag{3.21}$$

Thus we can obtain a particularly simple relationship between $\varphi(E)$ and q, the constant slowing down density. We next assume that we are below the threshold for inelastic scattering, and that only a single scattering material—normally a moderator—is present. (We may later modify the expression for combinations of materials.) Equation (2.47) provides the kernel for elastic scattering. Substituting it into Eq. (3.21) yields

$$\Sigma_s(E)\varphi(E) = \int_E^{E/\alpha} \frac{1}{(1-\alpha)E'}\Sigma_s(E')\varphi(E')dE'. \tag{3.22}$$

The solution may be shown to be

$$\Sigma_s(E)\varphi(E) = C/E \tag{3.23}$$

by simply inserting this expression into Eq. (3.22).

The normalization constant C is proportional to q, the number of neutrons slowed down by scattering past energy E. Examining Fig. 3.3, we observe that the number of neutrons that made their last scatter at E' ($>E$) to energies E'' ($<E$) will fall in the interval $\alpha E' \leq E'' \leq E$. Moreover, only neutrons with initial energies E' between E and E/α are capable of scattering to energies below E. Hence the number of neutrons slowing down past E per cm^3 in one second is

$$q = \int_E^{E/\alpha} \left[\int_{\alpha E'}^E \frac{1}{(1-\alpha)E'}\Sigma_s(E')\varphi(E')dE'' \right]dE'. \tag{3.24}$$

Substituting Eq. (3.23) for the flux, and performing the double integration, we obtain

$$q = \left[1 + \frac{\alpha}{1-\alpha}\ln\alpha \right]C. \tag{3.25}$$

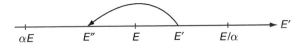

FIGURE 3.3   Energy loss from elastic scattering from energy E' to E''.

Noting that the bracketed term is identical to ξ, the slowing down decrement defined by Eq. (2.56), we may combine Eqs. (3.23) and (3.25) to represent the flux in terms of the slowing down density

$$\varphi(E) = \frac{q}{\xi \Sigma_s(E)E}. \tag{3.26}$$

This expression may be extended to situations where more than one scattering nuclide is present by adding their contributions to Eqs. (3.22) and (3.24). Suppose both fuel and moderator are present. Equation (3.26) still holds, where the scattering cross section becomes the sum over fuel and moderator, and the slowing down decrement is replaced by the weighted average defined by Eq. (2.61):

$$\bar{\xi} = \frac{\xi^f \Sigma_s^f(E) + \xi^m \Sigma_s^m(E)}{\Sigma_s^f(E) + \Sigma_s^m(E)}. \tag{3.27}$$

Between resonances the fuel and moderator scattering cross sections are nearly independent of energy. The flux is then proportional to $1/E$—and referred to as a "one-over-E" flux. Since the moderator is much lighter than the fuel, $\xi^f \ll \xi^m$, the fuel contribution to Eq. (3.26) is much less than that of the moderator.

Energy Self-Shielding

In the presence of resonance absorber, the flux is no longer proportional to $1/E$. However, we may obtain a rough estimate of its energy dependence by making some reasonable approximations. We assume that only fuel and moderator are present, and that only elastic scattering takes place. Equation (3.14) then reduces to

$$\Sigma_t(E)\varphi(E) = \int_E^{E/\alpha^f} \frac{1}{(1-\alpha^f)E'} \Sigma_s^f(E')\varphi(E')dE'$$
$$+ \int_E^{E/\alpha^m} \frac{1}{(1-\alpha^m)E'} \Sigma_s^m(E')\varphi(E')dE', \tag{3.28}$$

where for the energy range of resonance absorbers we have set $\chi(E) = 0$. Recall from Chapter 2 that a resonance is characterized by a width Γ. If the resonances are widely spaced, then the bulk of the resonance absorption will take place within about $\pm\Gamma$ of the resonance energy. Moreover, outside this interval absorption can be ignored and the flux approximated as $\propto 1/E$.

Scattering into the energy interval where absorption is most pronounced originates over a larger energy interval: between E and

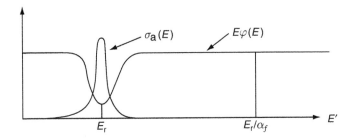

FIGURE 3.4 Energy loss from elastic scattering from energy to E'/E''.

$E + E/\alpha^f$ for the fuel, and between E and $E + E/\alpha^m$ for the moderator. In the narrow resonance approximation, which is valid for all but a few resonances, we assume both of these intervals to be much larger than the resonance width, as shown schematically in Fig. 3.4. In this case the preponderance of the areas under the integrals in Eq. (3.28) are occupied by the $1/E$ flux between resonances where absorption can be ignored and the scattering cross sections are energy independent. Thus we may insert Eq. (3.26) into the right side of Eq. (3.28) without much loss of accuracy. We evaluate the integrals with energy-independent constant cross section to obtain

$$\varphi(E) = \frac{q}{\bar{\xi}\Sigma_t(E)E},$$
(3.29)

where q is the neutron slowing down density above the resonance. Note that the only difference from Eq. (3.26) is that in the denominator the scattering has been replaced by the total cross section.

The total cross section, of course, includes both resonance absorption and scattering cross sections. Thus it increases greatly, causing the flux to decrease correspondingly, at energies where resonance absorption takes place. Such flux depression—illustrated in Fig. 3.4—is referred to as energy self-shielding. According to Eqs. (3.19) and (3.20), as neutrons slow down through a resonance the slowing down density is reduced by

$$\int \varphi(E)\Sigma_a(E)dE \approx \int \frac{\Sigma_a(E)}{\Sigma_t(E)E}\, dE\, \frac{q}{\bar{\xi}}.$$
(3.30)

Since self-shielding reduces the flux where the absorption cross section is large, it reduces overall neutron losses to absorption, and thus aids the propagation of the chain reaction. In the following chapter we will see that by lumping the fuel spatial self-shielding of the resonances serves further to reduce the absorption losses of neutrons. Other approximations can be applied when the resonance width is wider, but the qualitative effects of energy self-shielding remain the same.

Thermal Neutrons

At lower energies, in the thermal neutron range, we again use Eq. (3.15) as our starting point. The fission term on the right vanishes. The source of neutrons in this case comes from those scattering down from higher energies. We may represent this as a scattering source. We divide the integral in Eq. (3.15) according to whether E is less than or greater than the cutoff energy for the thermal neutron range, typically taken as $E_o = 1.0\,\text{eV}$. We then partition the equation as

$$\Sigma_t(E)\varphi(E) = \int_0^{E_o} \Sigma_s(E' \to E)\varphi(E')dE' + s(E)q_o, \quad E < E_o, \qquad (3.31)$$

where

$$s(E)q_o = \int_{E_o}^{\infty} \Sigma_s(E' \to E)\varphi(E')dE', \quad E < E_o, \qquad (3.32)$$

is just the source of thermal neutrons that arises from neutrons making a collision at energies $E' > E_o$, but having an energy of $E < E_o$ after that collision. The source may be shown to be proportional to the slowing down density at E_o, and if pure scattering and a $1/E$ flux is assumed at energies $E' > E_o$, a simple expression results for $s(E)$, the energy distribution of the source neutrons. In the thermal range the scattering distribution is difficult to represent in a straightforward manner, for not only thermal motion, but also binding of the target nuclei to molecules or within a crystal lattice must be factored into the analysis.

We may gain some insight by considering the idealized case of a purely scattering material. Then the solution of Eq. (3.31) would become time dependent, for without absorption in an infinite medium the neutron population would grow continuously with time since each slowed down neutron would go on scattering forever. If after some time the slowing down density were set equal to zero, an equilibrium distribution would be achieved satisfying the equation

$$\Sigma_s(E)\varphi_M(E) = \int_0^{E_o} \Sigma_s(E' \to E)\varphi_M(E')dE'. \qquad (3.33)$$

One of the great triumphs of kinetic theory was the proof that for this equation to be satisfied, the principle of detained balance must be obeyed. Detailed balance states that

$$\Sigma_s(E \rightarrow E')\varphi_M(E) = \Sigma_s(E' \rightarrow E)\varphi_M(E'), \qquad (3.34)$$

no matter what scattering law is applicable. Equally important, the principle states that in these circumstances the flux that satisfied this condition is the form found by multiplying the famed Maxwell-Boltzmann distribution, given by Eq. (2.34), by the neutron speed to obtain

$$\varphi_M(E) = \frac{1}{(kT)^2} E \exp(-E/kT) \qquad (3.35)$$

following normalization to

$$\int_0^\infty \varphi_M(E)dE = 1. \qquad (3.36)$$

In reality some absorption is always present. Absorption shifts the thermal neutron spectrum upward in energy from the Maxwell-Boltzmann distribution, since complete equilibrium is never reached before neutron absorption takes place. Figure 3.5 illustrates the upward shift, called spectral hardening, which increases with the size of the absorption cross section. Nevertheless, Eq. (3.35) provides a rough approximation to a reactor's thermal neutron distribution. A somewhat better fit to hardened spectra, such as those in Fig. 3.5, may be obtained by artificially increasing the temperature T by an amount that is proportional to $\Sigma_a/\xi\Sigma_s$.

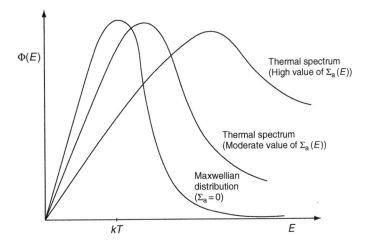

FIGURE 3.5 Thermal spectra compared to a Maxwell-Boltzmann distribution (adapted from A. F. Henry, *Nuclear-Reactor Analysis*, 1975, by permission of the MIT Press).

Fast and Thermal Reactor Spectra

Figure 3.6 shows typical neutron spectra plotted as $E\varphi(E)$ for a sodium-cooled fast reactor and for a water-cooled thermal reactor. Several features are noteworthy. Fast reactor spectra are concentrated in the keV and MeV range with nearly all of the neutrons absorbed before slowing down to energies less than a keV. Fast reactor cores contain intermediate weight elements, such as sodium coolant and iron used for structural purposes. These intermediate atomic weight elements have large resonances in their elastic scattering cross sections in the keV and MeV energy range. Thus the fast spectra are quite jagged in appearance, resulting from the energy self-shielding phenomenon, illustrated by Eqs. (3.16) and (3.29), in which the flux is inversely proportional to the total cross section.

Thermal reactor spectra have a more modest peak in the MeV range where fission neutrons are born. The spectra over higher energies are somewhat smoother as a result of the prominent role played by the lightweight moderator materials; moderators have no resonances at those energies, and therefore the cross sections in the denominators of Eqs. (3.16) and (3.29) are smoother functions of energy. Moving downward through the keV range, we see that the spectrum is nearly flat. Here there is very little absorption, resulting in a nearly $1/E$ [or constant $E\varphi(E)$] spectrum with the constant

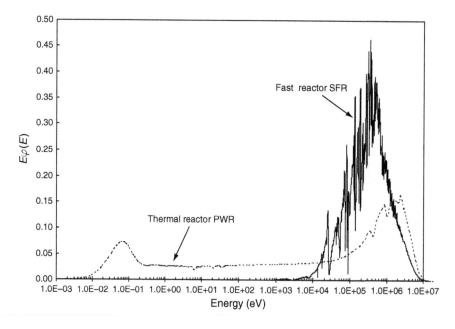

FIGURE 3.6   Neutron flux spectra from thermal (pressurized water) and fast (sodium-cooled) reactors (courtesy of W. S. Yang, Argonne National Laboratory).

slowing down density as given by Eq. (3.26). The thermal reactor spectra do decrease with decreasing energy going from 100 and 1.0 eV, accentuated by sharp dips in the flux. Although barely visible in the figure, resonance absorption in uranium over this energy range causes the slowing down density to decrease and the self-shielding indicated in Eq. (3.29) to become more pronounced. Below 1.0 eV, the characteristic thermal peak occurs. As a result of thermal neutron absorption in the fuel and moderator, the peak in the thermal spectrum is at an energy somewhat higher than would be indicated by the Maxwell-Boltzmann distribution given by Eq. (3.35). Finally, note that if we had plotted $\varphi(E)$ instead of $E\varphi(E)$ for the thermal reactor, the thermal flux peak would be millions of times larger than the peak of fission energy neutrons.

3.5 Energy-Averaged Reaction Rates

As the foregoing sections indicate, the ability to sustain a chain reaction depends a great deal on the distribution of neutrons in energy, which in turn is determined by the composition of nonfissile materials in the core and their effectiveness in slowing down the neutrons from fission toward thermal energies. To determine the overall characteristics of a reactor core, we must average cross sections and other data over the energy spectrum of neutrons. We accomplish this through the use of Eq. (3.13), which is termed the reaction rate for collisions of type x and has units of collisions/s/cm^3.

Reaction rates are commonly expressed as products of energy-averaged cross sections and the neutron flux:

$$\int_0^\infty \Sigma_x(E)\varphi(E)dE = \bar{\Sigma}_x\phi, \qquad (3.37)$$

where the cross section is

$$\bar{\Sigma}_x = \int_0^\infty \Sigma_x(E)\varphi(E)dE \bigg/ \int_0^\infty \varphi(E)dE, \qquad (3.38)$$

and the flux, integrated over energy, is

$$\phi = \int_0^\infty \varphi(E)dE. \qquad (3.39)$$

For a known neutron flux distribution, microscopic cross sections may also be averaged over energy. We simply make the replacement $\Sigma_x = N\sigma_x$ in Eqs. (3.37) and (3.38) to eliminate the atom density and obtain

$$\int_0^\infty \sigma_x(E)\varphi(E)dE = \bar{\sigma}_x\phi \tag{3.40}$$

and

$$\bar{\sigma}_x = \int_0^\infty \sigma_x(E)\varphi(E)dE \bigg/ \int_0^\infty \varphi(E)dE. \tag{3.41}$$

We may also express the flux as the product of the mean speed and the density of the neutrons:

$$\phi = \bar{v}n''', \tag{3.42}$$

where Eq. (3.9) defines the neutron density n'''. To accomplish this insert the flux definition given by Eq. (3.10) into Eq. (3.39):

$$\phi = \int_0^\infty v(E)\tilde{n}'''(E)dE \tag{3.43}$$

and note that to be consistent with Eq. (3.42) the mean speed must be defined by

$$\bar{v} = \int_0^\infty v(E)\tilde{n}'''(E)dE \bigg/ \int_0^\infty \tilde{n}'''(E)dE. \tag{3.44}$$

Frequently we will drop the bar indicating averaging from the left sides of Eqs. (3.38) and (3.41). Thus hereafter we assume that a cross section Σ_x or σ_x appearing without the (E) attached has been averaged over energy. If a cross section is independent of energy, we have $\sigma_x(E) \to \sigma_x$, and then, of course, we may take it outside the integral in Eqs. (3.38) and (3.41), and we have simply, $\bar{\sigma}_x = \sigma_x$ and $\bar{\Sigma}_x = \Sigma_x$.

More refined treatments of a neutron population often require cross section averaging over some limited range of neutron energies rather than over the entire neutron energy spectrum. The discussions of Section 3.4 indicate that the analysis of neutron spectra fall naturally into thermal, intermediate, and fast energy ranges. Correspondingly we may partition reaction rates as

$$\int \sigma_x(E)\varphi(E)dE = \int_T \sigma_x(E)\varphi(E)dE + \int_I \sigma_x(E)\varphi(E)dE$$
$$+ \int_F \sigma_x(E)\varphi(E)dE, \tag{3.45}$$

where hereafter attaching T, I, and F signifies integration over the ranges $0 \le E \le 1.0\,\text{eV}$, $1.0\,\text{eV} \le E \le 0.1\,\text{MeV}$, and $0.1\,\text{MeV} \le E \le \infty$, respectively. Employing Eq. (3.40), we may write this sum in terms of energy-averaged cross sections as

$$\bar{\sigma}_x \phi = \bar{\sigma}_{xT} \phi_T + \bar{\sigma}_{xI} \phi_I + \bar{\sigma}_{xF} \phi_F. \tag{3.46}$$

Each of the terms on the right results from multiplying and dividing the corresponding integral of Eq. (3.45) by

$$\phi_\Omega = \int_\Omega \varphi(E) dE, \quad \Omega = T, \, I, \, F, \tag{3.47}$$

and defining the energy averaged cross sections as

$$\bar{\sigma}_{x\Omega} = \int_\Omega \sigma_x(E) \varphi(E) dE \bigg/ \int_\Omega \varphi(E) dE, \quad \Omega = T, \, I, \, F. \tag{3.48}$$

More advanced so-called multigroup methods divide the energy spectrum into more than the three intervals shown here, and considerable effort is expended in determining the flux spectra in each group as accurately as possible. For our purposes, however, the division into thermal, intermediate, and fast energy segments is adequate. We perform cross section averaging by selecting appropriate flux approximations for use in Eqs. (3.46) through (3.48). We begin with the fast neutrons and work our way downward in energy.

Fast Cross Section Averages

Even though it includes only uncollided neutrons, Eq. (3.16) provides a first approximation to the flux distribution for fast neutrons. The total macroscopic cross section in the denominator, however, includes all of the nuclides present—fuel, coolant, and so on. Thus it is likely to be a strong and complex function of energy, particularly if significant concentrations of iron, sodium, or other elements that have scattering resonances in the MeV range are present. In Fig. 3.6 these effects are apparent in the jagged appearance of the fast flux for both thermal and fast reactors. To preclude the cross sections tabulated for individual elements from being dependent on the other elements present, we must further simplify Eq. (3.16) by taking $\Sigma_t(E)$ as energy independent. Then normalizing to $s_f''' / \Sigma_t = 1.0$, we have $\varphi(E) \approx \chi(E)$. Since only a very small fraction of fission neutrons are produced with energies less than 0.1 MeV we can extend the limits on the integrals in Eq. (3.48) from zero to infinity without loss of generality. With this proviso, the normalization condition of Eq. (2.32) sets the denominator equal to one, and Eq. (3.48) reduces to

$$\bar{\sigma}_{xF} = \int \sigma_x(E) \chi(E) dE. \tag{3.49}$$

Table 3.2 lists fast cross sections averaged over the fission spectrum for several of the isotopes that appear most prominently in

TABLE 3.2
Energy Averaged Microscopic Cross Sections (barns)

Nuclide	Thermal Spectrum Cross Sections			Resonance Integrals		Fast (Fission Spectrum) Cross Sections		
	σ_f	σ_a	σ_s	I_f	I_a	σ_f	σ_a	σ_s
^1_1H	0	0.295	47.7	0	0.149	0	3.92×10^{-5}	3.93
^2_1H	0	5.06×10^{-4}	5.37	0	2.28×10^{-4}	0	5.34×10^{-6}	2.55
$^{10}_5\text{B}$	0	3409	2.25	0	1722	0	0.491	2.12
$^{12}_6\text{C}$	0	3.00×10^{-3}	4.81	0	1.53×10^{-3}	0	1.23×10^{-3}	2.36
$^{16}_8\text{O}$	0	1.69×10^{-4}	4.01	0	8.53×10^{-5}	0	1.20×10^{-2}	2.76
$^{23}_{11}\text{Na}$	0	0.472	3.09	0	0.310	0	2.34×10^{-4}	3.13
$^{56}_{26}\text{Fe}$	0	2.29	11.3	0	1.32	0	9.22×10^{-3}	3.20
$^{91}_{40}\text{Zr}$	0	0.16	6.45	0	0.746	0	3.35×10^{-3}	5.89
$^{135}_{54}\text{Xe}$	0	2.64×10^6	—	0	7.65×10^3	0	7.43×10^{-4}	—
$^{149}_{62}\text{Sm}$	0	6.15×10^4	—	0	3.49×10^3	0	0.234	—

Table 3.2
(continued)

Nuclide	Thermal Spectrum Cross Sections			Resonance Integrals		Fast (Fission Spectrum) Cross Sections		
	σ_f	σ_a	σ_s	I_f	I_a	σ_f	σ_a	σ_s
$^{157}_{64}$Gd	0	1.92×10^5	1422	0	762	0	0.201	6.51
$^{232}_{90}$Th	0	6.54	11.8	0	84.9	7.13×10^{-2}	0.155	7.08
$^{233}_{92}$U	464	506	14.2	752	886	1.84	1.89	5.37
$^{235}_{92}$U	505	591	15.0	272	404	1.22	1.29	6.33
$^{238}_{92}$U	1.05×10^{-5}	2.42	9.37	2×10^{-3}	278	0.304	0.361	7.42
$^{239}_{94}$Pu	698	973	8.62	289	474	1.81	1.86	7.42
$^{240}_{94}$Pu	6.13×10^{-2}	263	1.39	3.74	8452	1.36	1.42	6.38
$^{241}_{94}$Pu	946	1273	11.0	571	740	1.62	1.83	6.24
$^{242}_{94}$Pu	1.30×10^{-2}	16.6	8.30	0.94	1117	1.14	1.22	6.62

Source: R. J. Perry and C. J. Dean, The WIMS9 Nuclear Data Library, Winfrith Technology Center Report ANSWERS/WIMS/TR.24, Sept. 2004.

power reactor cores. Such cross sections, however, provide only a smoothed approximation to what the cross sections would be if averaged over the actual flux distribution.

Resonance Cross Section Averages

Often the terms intermediate and resonance are used interchangeably in describing the energy range between 1.0 eV and 0.1 MeV because as neutrons slow down from fast to thermal energy the large cross sections caused by the resonances in uranium, plutonium, and other heavy elements account for the nearly all of the neutron absorption in this energy range. Equation (3.29) provides a reasonable approximation to the flux distribution in this energy range. However, as in the fast spectrum, the $\Sigma_t(E)$ term in the denominator is dependent on all of the constituents present in the reactor and thus must be eliminated in order to obtain cross sections that are independent of core composition. Ignoring the energy dependence of the total cross section, we simplify the flux to $\varphi(E) \approx 1/E$. Equation (3.48) then becomes

$$\bar{\sigma}_{xI} = \int_I \sigma_x(E) \frac{dE}{E} \bigg/ \int_I \frac{dE}{E}. \tag{3.50}$$

For capture and fission reactions intermediate range cross sections are frequently expressed as

$$\bar{\sigma}_{xI} = I_x \bigg/ \int_I \frac{dE}{E}, \tag{3.51}$$

where

$$I_x = \int \sigma_x(E) \frac{dE}{E} \tag{3.52}$$

defines the resonance integral. Since the predominate contributions to I_x $(x = a, f)$ arise from resonance peaks—such as those shown in Figs. 2.6, 2.9, and 2.10—that lie well within the range $1.0\,\text{eV} \leq E \leq 0.1\,\text{MeV}$, the values of resonance integrals are relatively insensitive to the limits of integration. The denominator of Eq. (3.51), however, depends strongly on those limits. Evaluating it between 1.0 eV and 0.1 MeV, then, yields

$$\bar{\sigma}_{xI} = 0.0869 I_x. \tag{3.53}$$

Table 3.2 includes the resonance integrals for common reactor constituents.

As the thermal reactor spectrum in Fig. 3.6 indicates, the $1/E$—that is, the $E\varphi(E) = \text{constant}$—spectrum is a reasonable approximation through the slowing down region. However, the dips that appear represent the resonance self-shielding that decreases the number of neutrons that are lost to absorption. Since Eq. (3.52) does *not* include the effects of self-shielding, numbers listed in Table 3.2 only provide an upper bound on resonance absorption, which would only be obtained in the limit of an infinitely dilute mixture of the resonance absorber in a purely scattering material. In reactor cores self-shielding dramatically reduces the amount of absorption. Advanced methods for calculating resonance absorption accurately are beyond the scope of this text. However, Chapter 4 includes empirical formulas that provide reasonable approximations to resonance absorption with the effects of self-shielding included.

Thermal Cross Section Averages

Although accurate determination of the thermal spectrum also requires advanced computational methods, averages over simplified spectra often serve as a reasonable first approximation in performing rudimentary reactor calculations. We approximate the thermal flux with the Maxwell-Boltzmann distribution, $\varphi(E) \approx \varphi_M(E)$, given by Eq. (3.35). With the normalization proved by Eq. (3.36), Eq. (3.48) thus reduces to

$$\bar{\sigma}_{xT} = \int \sigma_x(E)\varphi_M(E)dE. \tag{3.54}$$

Since $\varphi_M(E)$ is vanishingly small for energies greater than an electron volt, the upper limit on this integral can be increased from 1.0 eV to infinity without affecting its value. Thermal neutron cross sections averaged over the Maxwell-Boltzmann distribution at room temperature of 20 °C (i.e., 293 K) are tabulated for common reactor materials in Table 3.2. Appendix E provides a more comprehensive table of microscopic thermal cross sections integrated over the Maxwell-Boltzmann flux distribution as in Eq. (3.54), along with molecular weights and densities. The table includes all naturally occurring elements and some molecules relevant to reactor physics.

Frequently the cross sections are measured at 0.0253 eV, which corresponds to a neutron speed of 2200 m/s. The convention is based on what follows. The maximum—or most probable—value of $\varphi_M(E)$ may easily be shown to be

$$E = kT = 8.62 \times 10^{-5}T \text{ eV}, \tag{3.55}$$

where T is in degrees kelvin. We take the corresponding neutron speed to be

$$v = \sqrt{2E/m} = \sqrt{2kT/m} = 128\sqrt{T} \text{ m/s}. \qquad (3.56)$$

Cross section measurements made at $T_o = 293.61$ K yield $E_o = 0.0253$ eV and $v_o = 2,200$ m/s; 0.0253 eV and 2200 m/s are commonly referred to as the energy and speed of a thermal neutron. The cross sections tabulated in Table 3.2 and Appendix E, however, are the averages over the Maxwell-Boltzmann spectrum given by Eq. (3.54), rather than the cross sections evaluated at $E_o = 0.0253$ eV.

In the many cases where thermal scattering cross sections are independent of energy, Eq. (3.54) reduces to $\bar{\sigma}_{sT} = \sigma_s$. At thermal energies, however, the binding of atoms to molecules or within crystal lattices can significantly affect the thermal scattering cross sections. To account for this, the cross sections for hydrogen, deuterium, and carbon given in Table 3.2 and Appendix E are corrected to include the effects of such binding. These corrections allow Eqs. (2.14) and (2.15), for example, to be used without modification in the determination of the thermal scattering cross section of water.

In contrast, many thermal absorption cross sections are proportional to $1/v$:

$$\sigma_a(E) = \sqrt{E_o/E}\, \sigma_a(E_o). \qquad (3.57)$$

To obtain the energy-averaged cross section in such cases we must substitute this equation and Eq. (3.35) into Eq. (3.54):

$$\bar{\sigma}_{aT} = \int_0^\infty \sqrt{E_o/E}\, \sigma_a(E_o) \frac{1}{(kT)^2} E \exp(-E/kT) dE. \qquad (3.58)$$

Evaluating the integral, we obtain

$$\bar{\sigma}_{aT} = \frac{\sqrt{\pi}}{2}\left(\frac{E_o}{kT}\right)^{1/2} \sigma_a(E_o) = 0.8862(T_o/T)^{1/2}\sigma_a(E_o). \qquad (3.59)$$

Thus the $1/v$ absorption cross section is dependent on the absolute temperature, and even if $T = T_0$ the averaged absorption cross sections are not the same as those measured at E_o. In Table 3.2 and Appendix E, thermal absorption and capture cross sections are the averages defined by $\bar{\sigma}_{aT}$ of Eq. (3.54).

To correct these $1/v$ thermal cross sections for temperature, we note from Eq. (3.59) that $\bar{\sigma}_{aT}(T) = (T_o/T)^{1/2}\bar{\sigma}_{aT}(T_o)$. In dealing with macroscopic thermal cross sections correcting for temperature becomes more complex if the material has a significant coefficient of thermal expansion. Since $\Sigma_x = N\sigma_x$, and the atom density is given by $N = \rho N_0/A$, density decreases with increasing temperature will also cause macroscopic cross sections to decease even if the microscopic cross sections remain constant.

3.6 Infinite Medium Multiplication

We conclude this chapter by returning to the calculation of the multiplication, k_∞, the ratio of the number of fission neutrons produced to the number of neutrons absorbed. The ratio is determined by using the reaction rate definition, Eq. (3.13). Since the number of fission neutrons produced is $\int_0^\infty \nu\Sigma_f(E)\varphi(E)dE$, where ν is the number of neutrons/fission, and the number of neutrons absorbed is $\int_0^\infty \Sigma_a(E)\varphi(E)dE$, we have

$$ k_\infty = \int_0^\infty \nu\Sigma_f(E)\varphi(E)dE \bigg/ \int_0^\infty \Sigma_a(E)\varphi(E)dE. \qquad (3.60) $$

Using the definitions of the energy-averaged cross sections and flux given in Eqs. (3.37) we may express k_∞ as a ratio of cross sections:

$$ k_\infty = \nu\bar{\Sigma}_f/\bar{\Sigma}_a, \qquad (3.61) $$

where only fissionable materials contribute to the numerator, while absorption cross sections of all of the reactor core's constituents contribute to the denominator.

Thus far we have assumed implicitly that the fuel, moderator, coolant, and other core constituents are all exposed to the same energy-dependent flux $\varphi(E)$. Provided the volumes of core constituents are finely mixed—for example, powders of uranium and graphite—this assumption holds. However, in power reactors the diameters of fuel elements, the spacing of coolant channels, and geometric configurations of other constituents result in larger separations between materials. In these circumstances the flux magnitudes to which the fuel, coolant, and/or moderator are exposed often are not identical. Power reactor cores consist of lattices of cells, each consisting of a fuel element, coolant channel, and in some cases a separate moderator region. The expressions derived above remain valid provided we interpret them as spatial averages over the constituents of one such cell, with account taken for differences in flux magnitudes. In the following chapter we first examine the lattice structures of power reactors. We then take up the modeling of fast and thermal reactor lattices in order to examine these differences in flux magnitudes and then to obtain expressions for k_∞ explicitly in terms of the various core constituents.

Bibliography

Cullen, D. E., "Nuclear Cross Section Preparation," *CRC Handbook of Nuclear Reactor Calculations, I,* Y. Ronen, ed., CRC Press, Boca Raton, FL, 1986.

Duderstadt, James J., and Louis J. Hamilton, *Nuclear Reactor Analysis,* Wiley, NY, 1976.

Henry, Allen F., *Nuclear-Reactor Analysis,* MIT Press, Cambridge, MA, 1975.

Jakeman, D., *Physics of Nuclear Reactors,* Elsevier, NY, 1966.

Mughabghab, S. F., *Atlas of Neutron Resonances,* Elsevier, Amsterdam, 2006.

Stacey, Weston M., *Nuclear Reactor Physics,* Wiley, NY, 2001.

Templin, L. J., ed., *Reactor Physics Constants,* 2nd ed., ANL-5,800, Argonne National Laboratory, Argonne, IL, 1963.

Williams, M. M. R., *The Slowing Down and Thermalization of Neutrons,* North-Holland, Amsterdam, 1966.

Problems

3.1. Verify Eqs. (3.23) and (3.25).

3.2. Show that in Eq. (3.31) the normalization condition $\int_0^{E_o} s(E)dE = 1$ must be obeyed. Hint: Note that $\int_0^{E_o} p(E' \to E)dE = 1$ for $E' \leq E_o$.

3.3. In Eq. (3.31) suppose that the neutron slowing down past E_o is due entirely to elastic scattering from a single nuclide with $A > 1$, and with no absorption for $E > E_o$. Show that $s(E)$ then takes the form

$$s(E) = \begin{cases} \dfrac{1}{(1-\alpha)\xi}\left(\dfrac{1}{E_o} - \dfrac{\alpha}{E}\right), & \alpha E_o < E < E_o, \\ 0, & E < \alpha E_o \end{cases}$$

3.4. For thermal neutrons calculate $\bar{\eta}$ as a function of uranium enrichment and plot your results. Use the uranium data from the following table:

	ν	σ_f (barns)	σ_a (barns)
Uranium-235	2.43	505	591
Plutonium-239	2.90	698	973
Uranium-238	—	0	2.42

3.5. Suppose a new isotope is discovered with a "$1/E$" absorption cross section given by $\Sigma_a(E) = (E_o/E)\Sigma_a(E_0)$. Determine the energy-averaged cross section if the isotope is placed in the thermal flux distribution given by Eq. (3.35).

3.6. In the wide resonance approximation (also called narrow resonance infinite mass approximation because the fuel is assumed to have an infinite mass), $A^f \to \infty$ and thus $\alpha^f \to 1$ in the first integral on the right of Eq. (3.28) while the remaining approximations are the same as in narrow resonance approximation. Determine $\varphi(E)$ through the resonance. How does it differ from Eq. (3.29)? In which case is there more energy self-shielding?

3.7. Lethargy defined as $u = \ln(E_o/E)$ is often used in neutron slowing down problems; lethargy increases as energy decreases. Note the following transformations: $\varphi(E)dE = -\varphi(u)du$, $p(E \to E') dE' = -p(u \to u')du'$, and $\Sigma_x(E) = \Sigma_x(u)$.

a. Show that $p(E \to E')$ given by Eq. (2.47) becomes

$$p(u \to u') = \begin{cases} \dfrac{1}{1-\alpha}\exp(u - u'), & u \le u' \le u + \ln(1/\alpha), \\ \\ 0, & \text{otherwise} \end{cases}$$

b. Express Eq. (3.22) in terms of u.

3.8. Making a change of variables from energy to speed, show that Eq. (2.47) becomes

$$p(v \to v') = \begin{cases} \dfrac{2v'}{(1-\alpha)v^2}, & v\sqrt{\alpha} \le v' \le v, \\ \\ 0, & \text{otherwise} \end{cases}$$

3.9. Suppose that the Maxwell-Boltzmann distribution, Eq. (2.34), represents the neutron density in Eqs. (3.43) and (3.44):

a. Find the value of $\bar{v}$.
b. If we define $\bar{E} \equiv \frac{1}{2}m\bar{v}^2$, show that $\bar{E} = 1.273kT$.
c. Why is your result different from the average energy of $\frac{3}{2}kT$ given by Eq. (2.33)?

3.10. A power reactor is cooled by heavy water (D_2O) but a leak causes a 1.0 atom % contamination of the coolant with light water (H_2O). Determine the resulting percentage increase or decrease in the following characteristics of the coolant:

a. Slowing down decrement.
b. Slowing down power.
c. Slowing down ratio.

3.11. Using the data in Appendix E calculate the microscopic absorption cross section of water, averaged over a thermal neutron spectrum:

a. At room temperature.
b. At 300 °C, which is a typical operating temperature for a water-cooled reactor.

3.12. Repeat problem 3.11 for heavy water.

CHAPTER 4

The Power Reactor Core

4.1 Introduction

Two criteria play dominant roles in determining the composition of a power reactor core: Criticality must be maintained over the range of required power levels and over the life of the core as fuel is depleted. The design must also allow the thermal energy produced from fission to be transferred out of the core without overheating any of its constituents. Many other considerations also come into play: the mechanical support of the core structure, stability and control of the chain reaction under widely varied circumstances, and so on. But the neutron physics discussed in the preceding chapters and heat transfer interact most strongly in determining the construction of power reactor cores. This chapter first examines the core layouts of the more common classes of power reactors, relating heat transport and neutronic behavior. It then presents in more detail the impact of the reactor lattice structures on the neutronic behavior and in particular on the determination of the multiplication.

4.2 Core Composition

Reactors have been designed with a wide variety of configurations. These include cores consisting of molten materials in a container, in which the liquid fuel itself is piped through circulation loops from which the heat is removed, and pebble bed reactors where the fuel consists of a bed of solid spheres through which coolant is circulated to remove the heat. Most power reactors, however, are cylindrical in shape with coolant flowing through channels extending the axial length of the core. These channels are one constituent of a periodic lattice consisting of cylindrical fuel elements, coolant channels, and in some thermal reactors a separate moderator region.

Figure 4.1 illustrates the fuel–coolant–moderator lattice structure for four diverse classes of power reactors. In all cases, heat from fission is produced within the fuel and conducted to the coolant

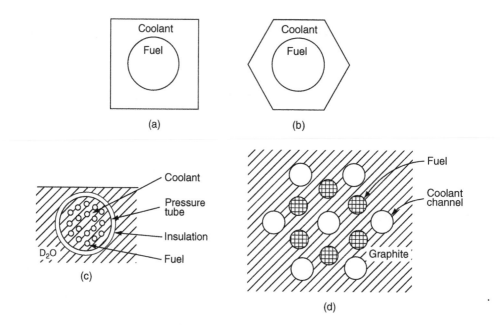

FIGURE 4.1 Reactor lattice cross sections (not on the same scale). (a) Water-cooled reactor, (b) fast reactor, (c) CANDU heavy water reactor, (d) high temperature gas-cooled graphite reactor.

channel surface. It then is convected into the coolant and transported axially along the coolant channel and out of the core. The four diagrams that make up Fig. 4.1, however, are drawn on quite different scales, since in all cases the fuel element diameters are of the order of 1 cm. Thermal constraints on the heat flux crossing a fuel element's surface and on the temperature along its centerline limit both its diameter and the power per unit length—called the linear heat rate or q'—that it can produce. Since allowable linear heat rates typically fall within the range between few and tens of kW/m, a large power reactor designed to produce 1000 MW(t) or more of heat must contain many thousands of cylindrical fuel elements—often referred to as fuel pins.

Refueling a reactor containing thousands of fuel elements by replacing them one by one would represent an inordinately time-consuming and hence uneconomical task. Thus fuel elements are grouped together to form fuel assemblies. The mechanical design of fuel assemblies allows them be moved as a whole in and out of the reactor during refueling procedures. Figure 4.2 illustrates three examples of fuel assemblies. The assemblies' cross-sectional areas may be square or hexagonal as shown in Figs. 4.2a and 4.2c. Fuel elements may also be bundled into circular fuel assembles as shown in Fig. 4.2b; in this case the bundles are inserted in tubes placed in

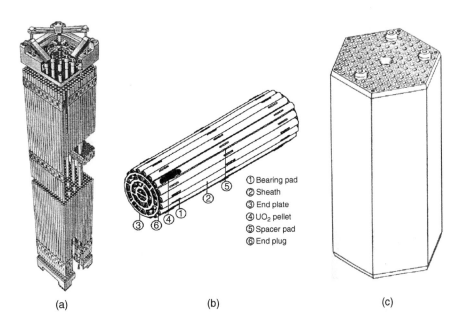

① Bearing pad
② Sheath
③ End plate
④ UO₂ pellet
⑤ Spacer pad
⑥ End plug

(a)   (b)   (c)

FIGURE 4.2 Reactor fuel assemblies. (a) Pressurized water reactor (courtesy of Westinghouse Electric Company), (b) CANDU heavy water reactor (courtesy of Atomic Energy of Canada, Ltd.), (c) high temperature gas-cooled reactor (courtesy of General Atomics Company).

square or hexagonal arrays within moderator regions as indicated, for example, in Fig. 4.1c.

Figure 4.3 depicts lateral cross sections of a power reactor cores made up of square and hexagonal fuel assemblies, respectively. The shading in Fig. 4.3 indicates that typically not all of the fuel assemblies in a reactor core are identical. They may differ in fuel enrichment in order to flatten the power across the core, or they may have been placed in the core during different refueling operations. The placement of control poisons may also cause assemblies to differ.

In addition to fuel, coolant, and (in thermal reactors) moderator, the reactor must contain channels placed at carefully designated intervals throughout such lattice configurations to allow the insertion of control rods. These rods consist of strong neutron absorbers—often referred to as neutron poisons—such as boron, cadmium, or hafnium. Their insertion controls the reactor multiplication during power operations, and they shut down the chain reaction when fully inserted. Some classes of power reactors contain space for the control rods as channels reserved for them within designated fuel assemblies; the assemblies shown in Figs. 4.2a and 4.2c fall into this category. In other systems the control rods are inserted between the fuel assemblies. For example, control rods with cruciform cross sections may be

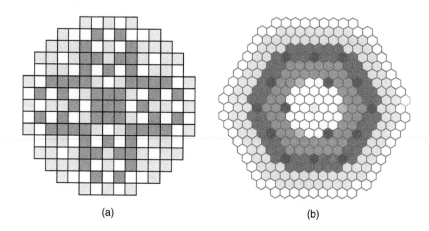

(a) (b)

FIGURE 4.3 Reactor cores consisting of square and hexagonal fuel assemblies (courtesy of W. S. Yang, Argonne National Laboratory). (a) Square fuel assemblies, (b) hexagonal fuel assemblies.

placed at the intersections of square assemblies, or control rods may be inserted into moderator regions between assemblies.

The fuel elements' linear heat rate and the volume ratio of coolant and/or moderator to fuel, designated as V_x/V_f, along with a number of other factors determine $\bar{P}'''$, the average power density, that is, the average power produced per unit volume, achievable for each reactor class. Since a reactor's power is given by $P = \bar{P}'''V$, for a given power the core volume V, it is inversely proportional to the average power density. Core lattice structures must be optimized to facilitate heat transfer, and thus maximize achievable power densities within the heat removal capabilities of the coolant. But neutronic considerations play an equally important role. For lattice structure and particularly the ratios of fuel, coolant, and moderator largely determine the value of k_∞ for a given fuel enrichment. On the other hand, P_{NL}, the non-leakage probability, approaches a value of one as the core volume increases. Thus since $k = k_\infty P_{NL}$, lattice structure and achievable power density intertwine in determining the critical state of a power reactor.

Table 4.1 compares typical parameters for some major classes of power reactors. To gain a better understanding of how the displayed values result from power reactor design and operational considerations we next examine separately each broad class of the reactors in terms of the neutronic properties of their fuel, moderator, and/or coolant.

Light Water Reactors

Pressurized water reactors (PWRs) and boiling water reactors (BWRs) utilize ordinary water both as coolant and moderator. Both these

TABLE 4.1
Representative Reactor Lattice Properties

	PWR Pressurized-H_2O Reactor	BWR Boiling-H_2O Reactor	PHWR CANDU-D_2O Reactor	HTGR C-Moderated Reactor	SFR Na-Cooled Fast Reactor	GCFR He-Cooled Fast Reactor
$\bar{q}'$ (kW/m) average linear heat rate	17.5	20.7	24.7	3.7	22.9	17.0
V_x/V_f volume ratio[a]	1.95	2.78	17	135	1.25	1.93
$\bar{P}''$ (MW/m³) average power density	102	56	7.7	6.6	217	115
V (m³) volume 3000 MW(t) reactor	29.4	53.7	390	455	13.8	26.1
$\tilde{e}$ (weight %) enrichment	4.2	4.2	0.7	15	19	19

[a] x = moderator for a thermal reactor and coolant for a fast reactor, f = fuel.
Source: Data courtesy of W. S. Yang, Argonne National Laboratory.

classes of light water reactors (LWRs) utilize square lattice cells, similar to that shown in Fig. 4.1a. The fuel consists of uranium dioxide pellets clad in zirconium for structural support and to prevent fission product leakage into the coolant. Moderator to fuel volume ratios are roughly 2:1, which is near optimal for maximizing k_∞ in water-moderated systems. The optimal volume ratio derives from the water's neutron slowing down properties listed in Table 3.1. Water has the largest slowing down power of any moderator as a result of the small mass and substantial scattering cross section of hydrogen. However, water also has the largest thermal absorption cross section of any of the listed moderators, which manifests itself as the smallest value of the slowing down ratio. Thus although water is excellent for slowing down neutrons, if larger ratios of water to fuel volume are employed in reactor lattices, the increased thermal absorption in the moderator results in an unacceptable decrease in k_∞. Water's combination of having the largest slowing down power but the smallest slowing down ratio of any of the moderators leads to LWR designs with substantially smaller moderator to fuel volume ratios than found in heavy water or graphite-moderated reactors.

These smaller ratios—and the fact that the water serves both as moderator and coolant—leads to PWRs and BWRs having the most compact lattices of any of the thermal reactors. As Table 4.1 indicates, the comparatively small moderator to fuel volume ratio leads to higher power densities and smaller reactor volumes than for heavy water reactors or graphite-moderated systems of the same power. Water's large thermal absorption cross section, however, precludes the possibility of achieving criticality with natural uranium fuel in a LWR. To overcome this obstacle water-moderated reactors must employ slightly enriched fuel, typically in the range between 2% and 5%.

In PWRs, the core is contained in a vessel pressurized to 1520 bar (~2200 psi) to prevent coolant boiling at operating temperatures in the range of 316 °C (~600 °F). As Fig. 4.4a indicates, water exiting the core circulates through heat exchangers, called steam generators, before being pumped back to the core inlet. The secondary side of the steam generator operates at a lower pressure such that feed water entering it boils, thus supplying steam to the turbine. Coolant temperatures are similar for BWRs. However, they operate at less elevated pressures of 690 bar (~1000 psi), allowing boiling to take place in the coolant channels. As Fig. 4.4b indicates, in BWRs, the feed water enters directly into the reactor vessel, and the steam generated within the reactor passes directly to the turbine, eliminating the need for steam generators.

Water-moderated reactors operate in batch mode fuel cycles: The reactor is shut down at regular intervals, ranging from 1 to 2 years. During shutdown, typically lasting a number of weeks, 20 to 30% of the assemblies containing fuel from which the fissile material is

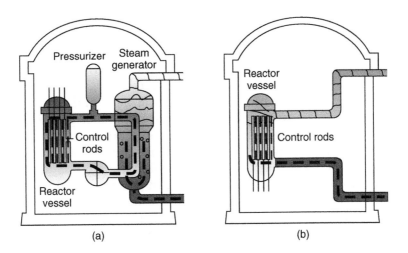

FIGURE 4.4   Light water cooled reactors (courtesy U.S. Nuclear Regulatory Commission). (a) Pressurized water reactor, (b) boiling water reactor.

most depleted are removed and replaced by fresh assemblies. In batch mode operation the fuel must be sufficiently enriched to compensate for the uranium burn up that takes place over the length of time that the assembly remains in the reactor.

During core life neutron poisons control the reactor multiplication. In PWRs soluble boron in the coolant serves this purpose, and in most thermal reactors burnable poisons placed in the fuel or elsewhere also serve to compensate for fuel burn up. Control rods must be present to rapidly shut down the chain reaction, but they may also be used to compensate for fuel depletion. As the schematic diagrams of Fig. 4.4 indicate, PWR control rods are inserted from the top, whereas those of BWRs are inserted from the bottom. The PWR control rods are employed in clusters that occupy channels within the fuel assemblies as depicted in Fig. 4.2a. In BWRs the control rods have cruciform-shaped cross sections and fit into channels between the square fuel assemblies, with their centerlines positioned at intersections between four assemblies.

Heavy Water Reactors

As Table 3.1 indicates, the larger mass of deuterium causes heavy water's slowing down power to be substantially smaller than that of H_2O. However, deuterium's thermal absorption cross section is minuscule, and as a result D_2O has the largest slowing down ratio of any moderator. Thus in contrast to LWRs, reactors moderated by heavy water require large volume ratios of moderator to fuel to provide adequate neutron slowing down. At the same time they can tolerate larger moderator volumes because the thermal absorption cross section of deuterium is so small.

Pressurized heavy water reactors (PHWRs) of the CANDU type are by far the most common D_2O-moderated power systems today. A CANDU core consists of a large cylindrical tank, called a calandria, placed on its side; an array of horizontal pressure tubes passes through the calandria. Each pressure tube houses several fuel assembly segments, often called bundles, each containing 30 to 40 fuel pins as pictured in Figs. 4.1c and 4.2b. The fuel pins are similar to those in LWRs. They consist of UO_2 pellets, clad in zirconium. Heavy water, which also serves as coolant, is pumped through the tubes, circulated through steam generators similar in design to those found in PWRs, and returned to the core inlet. The substantial spacing between pressure tubes within the D_2O-filled calandria provides the reactor's large moderator to fuel volume ratio. The calandria is insulated from the pressure tubes, allowing it to operate near atmospheric pressure and room temperature. Thus only the tubes containing the fuel assemblies need be maintained at high enough pressure to keep the smaller volume of D_2O coolant from boiling at operating temperatures.

The moderator properties of heavy water—most particularly its small thermal absorption cross section—allow PHWRs to be fueled with natural uranium. However, without enrichment the fuel cannot sustain the levels of burn up required to operate the reactor for a year or more between refueling. Instead, CANDU reactors undergo continuous refueling while they are operating: A pair of on-line refueling machines isolate one pressure tube at a time and insert a fresh fuel bundle into one end of the core while removing a depleted one from the other. This mode of operation requires less control poison to maintain the reactor in a critical state than would be required with batch refueling. The control rods, which pass through the calandria, outside of the pressure tubes, serve primarily for reactor shutdown.

Graphite-Moderated Reactors

Table 3.1 indicates that the heavier atomic mass of carbon causes graphite's slowing down power to be smaller than that of either light or heavy water. Its thermal absorption cross section, however, is small enough to give it a slowing down ratio intermediate between light and heavy water. The net effect is that graphite reactor lattices designed to maximize k_∞ have very large moderator to fuel volume ratios. A number of early power reactor designs were carbon dioxide–cooled graphite-moderated systems that utilized natural uranium fuel. Graphite's smaller values of the slowing down power and ratio made the task more difficult than with D_2O moderator. Consequently as partially enriched fuel became available it replaced natural uranium in subsequent designs. More recently, helium coolant combined with partially enriched fuel has led to the design of graphite-moderated power reactors capable of operating at very high temperatures.

Table 4.1 displays parameters for a high temperature gas-cooled reactor (HTGR). The core contains only graphite and ceramic materials, allowing the helium coolant to achieve higher coolant temperatures than if metal cladding or structures were present. The lattice structure is similar to that shown in Fig. 4.1d; heat generated in the fuel passes through the graphite moderator before being carried away by the gas coolant. The fuel consist of compacts of uranium carbide particles compacted in graphite, thus further increasing the moderator to fuel ratio. The fuel enrichment is quite high. Each fuel assembly, as pictured in Fig. 4.2c, consists of a prismatic block of graphite containing two arrays of holes, for fuel and coolant, respectively. The control rods occupy other axial holes in the prismatic graphite block. The helium coolant circulates through a steam generator or gas turbine and back into the core. Refueling of HTGRs is accomplished in batch mode, similar to that employed with water-cooled reactors. The large volume of graphite moderator required causes the HTGR to have the lowest power density and thus the largest volume of any of the reactors listed in Table 4.1

RBMK Reactors

Thus far we have examined thermal reactors either in which the coolant and moderator are the same or in which the coolant is a gas. In the latter case the gas density is sufficiently small that the volume which it occupies has relatively minor effects on the determination of k_∞. Other power reactors have been based on using liquid coolants that differ from the moderator. For example, designs similar to the CANDU systems may employ pressurized or boiling light water coolant in conjunction with heavy water moderator. Less conventional is the molten salt reactor in which the molten fuel also serves as coolant by circulating through a core that consists of a fixed graphite moderator structure. Other coolant–moderator combinations have been employed, but for the most part only in prototypical reactors. In contrast the Russian RBMK reactor design has been widely used for power production throughout the former Soviet Union.

The RBMK is a water-cooled, graphite-moderated reactor. The requirement for large volumes of moderator translates into low power densities and volumes as large as $1000\,m^3$. The RBMK has some similarities to CANDU reactor designs in that the core consist of pressure tubes which contain fuel assemblies composed of bundles of cylindrical fuel elements of uranium dioxide in zirconium cladding. The pressure tubes carry the water coolant through the graphite moderator, which forms the bulk of the core volume. Like the CANDU reactors, the RBMK refuels on-line using machines to isolate one pressure tube at a time. The differences between RBMK and CANDU designs, however, are great. The RBMK design's pressure tubes transverse the core vertically, and carry light water coolant which boils as it passes through the tubes. The tubes as well as control rods occupy holes in the large bocks of graphite moderator, and finally the fuel is enriched to approximately 2%.

Fast Reactors

Fast reactor cores contain as little low atomic mass material as possible in order to impede neutron slowing down by elastic scattering. Even then, fast reactors require enrichments of 10%, or more. Hexagonal lattice cells similar to that pictured in Fig. 4.1b are employed because smaller volume ratios of coolant to fuel can be utilized than in square lattices. As Table 4.1 indicates, these tightly packed lattices are instrumental in causing fast reactors to achieve higher power densities, and hence smaller volumes, than are found in thermal reactors.

Fast reactor fuel may be metal or a ceramic, encapsulated in metal cladding. Liquid metals are the most widely used coolant because their atomic weight is larger than other liquids, and they have excellent heat transfer properties and can be employed in low-pressure systems. Sodium-cooled fast reactors (SFRs) are the most common designs. Because sodium reacts violently with water, however, SFRs require the placement of an intermediate heat exchanger between the reactor core and the steam generator; should a steam generator leak occur, the sodium that passes through the reactor would not make contact with water. Some Russian fast reactors have utilized molten lead coolant. Gas-cooled fast reactors (GCFRs) offer an alternative to liquid metal–cooled systems since the gas's low density causes it to have no appreciable effect on the neutron spectrum. However, high pressure and large rises in the coolant temperatures are then required to achieve adequate heat transport. Like most other systems, fast reactor refueling is carried out in batch mode.

4.3   Fast Reactor Lattices

In the preceding chapter we examined neutron spectra assuming that all the constituents of a reactor core were exposed to the same energy dependent flux $\varphi(E)$. The chapter concluded using energy-averaged cross sections to express the multiplication as

$$k_\infty = \int_0^\infty \nu\Sigma_f(E)\varphi(E)dE \bigg/ \int_0^\infty \Sigma_a(E)\varphi(E)dE, \qquad (4.1)$$

which serves as the starting point for the more detailed analysis of both fast and thermal reactors. Fast reactors differ from thermal reactors in several respects. The fuel enrichments are higher than typically found in thermal reactors, generally substantially exceeding 10%. Core designers eliminate low atomic weight materials to the greatest extent possible, since they have the adverse effect of degrading the neutron energy spectrum. The result is a fast neutron spectrum such as that shown in Fig. 3.6. Since cross sections in general decrease with increasing neutron energy, the neutron spectrum averaged cross sections in a fast reactor are substantially smaller than those in a thermal system. Accordingly, in fast reactors neither the fuel diameter nor the coolant thickness between fuel pins substantially exceeds a mean free path. Under such conditions the spatial distribution of the flux will be quite flat across the lateral cross section of the lattice cell, allowing us to approximate that fuel, coolant, and any structural material are all exposed to the same flux distribution, $\varphi(E)$. Thus for any reaction x, we may use Eq. (2.18) to volume-weight the cross sections of fuel, coolant and structural materials:

$$\Sigma_x(E) = (V_f/V)\Sigma_x^f(E) + (V_c/V)\Sigma_x^c(E) + (V_{st}/V)\Sigma_x^{st}(E), \qquad (4.2)$$

where the cell volume is the sum of contributions from these three components: $V = V_f + V_c + V_{st}$. Substituting Eq. (4.2) into Eq. (4.1) yields

$$k_\infty = \frac{V_f \int_0^\infty \nu\Sigma_f^f(E)\varphi(E)dE}{V_f \int_0^\infty \Sigma_a^f(E)\varphi(E)dE + V_c \int_0^\infty \Sigma_a^c(E)\varphi(E)dE + V_{st} \int_0^\infty \Sigma_a^{st}(E)\varphi(E)dE},$$

$$(4.3)$$

where only the fuel contributes to the fission cross section in the numerator.

We may express the multiplication in terms of energy-averaged (also referred to as one-energy-group) cross sections as follows. First integrate the flux over energy:

$$\phi = \int_0^\infty \varphi(E)dE. \qquad (4.4)$$

Then defining the flux-averaged cross section as in Eq. (3.38), we have

$$\bar{\Sigma}_x^y = \int_0^\infty \Sigma_x^y(E)\varphi(E)dE \bigg/ \int_0^\infty \varphi(E)dE, \tag{4.5}$$

where y indicates the material in which the reaction is taking place. Combining these two equations then expresses the reaction rate as the product of the energy-averaged cross section and the flux:

$$\int_0^\infty \Sigma_x^y(E)\varphi(E)dE = \bar{\Sigma}_x^y \phi. \tag{4.6}$$

The cell-averaged reaction rate then becomes

$$\int_0^\infty \Sigma_x(E)\varphi(E)dE = \left(\frac{V_f}{V}\bar{\Sigma}_x^f + \frac{V_c}{V}\bar{\Sigma}_x^c + \frac{V_{st}}{V}\bar{\Sigma}_x^{st}\right)\phi, \tag{4.7}$$

and we may write Eq. (4.3) as

$$k_\infty = \frac{V_f\nu\bar{\Sigma}_f^f}{V_f\bar{\Sigma}_a^f + V_c\bar{\Sigma}_a^c + V_{st}\bar{\Sigma}_a^{st}}. \tag{4.8}$$

The fuel enrichment and the ratios of coolant and other materials to fuel nuclei become the primary determinants of the lattice multiplication. To examine enrichment we write the number density of fuel atoms as a sum of the fissile (*fi*) and fertile (*fe*) contributions:

$$N_f = N_{fi} + N_{fe}. \tag{4.9}$$

We define enrichment $\tilde{e}$ as in Eq. (3.6) to be the ratio of fissile to total fuel nuclei:

$$\tilde{e} = N_{fi}/N_f. \tag{4.10}$$

We next use Eq. (2.5) to specify macroscopic cross sections (for reaction x in material y) in terms of their microscopic counterparts,

$$\bar{\Sigma}_x^y = N_y\bar{\sigma}_x^y, \tag{4.11}$$

where the energy-averaged microscopic cross sections are

$$\bar{\sigma}_x^y = \int_0^\infty \sigma_x^y(E)\varphi(E)dE \Big/ \int_0^\infty \varphi(E)dE. \qquad (4.12)$$

Separating the fuel into fissile and fertile contributions reduces Eq. (4.11) to

$$\bar{\Sigma}_x^f = N_{fi}\bar{\sigma}_x^{fi} + N_{fe}\bar{\sigma}_x^{fe}. \qquad (4.13)$$

Thus utilizing Eqs. (4.9) through (4.11) in this expression allows us to obtain the microscopic fuel cross section as

$$\bar{\sigma}_x^f = \tilde{e}\bar{\sigma}_x^{fi} + (1 - \tilde{e})\bar{\sigma}_x^{fe}. \qquad (4.14)$$

The foregoing definitions allow us to express k_∞ in terms the enrichment and these microscopic cross sections. Thus Eq. (4.8) becomes

$$k_\infty = \frac{V_f N_f \left[\tilde{e}\nu^{fi}\bar{\sigma}_f^{fi} + (1 - \tilde{e})\nu^{fe}\bar{\sigma}_f^{fe}\right]}{V_f N_f \left[\tilde{e}\bar{\sigma}_a^{fi} + (1 - \tilde{e})\bar{\sigma}_a^{fe}\right] + V_c N_c \bar{\sigma}_a^c + V_{st} N_{st} \bar{\sigma}_a^{st}}, \qquad (4.15)$$

or alternatively

$$k_\infty = \frac{\tilde{e}\nu^{fi}\bar{\sigma}_f^{fi} + (1 - \tilde{e})\nu^{fe}\bar{\sigma}_f^{fe}}{\tilde{e}\bar{\sigma}_a^{fi} + (1 - \tilde{e})\bar{\sigma}_a^{fe} + (V_c N_c/V_f N_f)\bar{\sigma}_a^c + (V_{st} N_{st}/V_f N_f)\bar{\sigma}_a^{st}}. \qquad (4.16)$$

Because the ratio of $\nu\sigma_f$ to σ_a is larger for fissile than for fertile materials, fast reactor multiplication increases with enrichment. The effects of the coolant and structural materials are subtler. As Eq. (4.16) indicates, increasing the ratio of coolant to fuel atoms (i.e., increasing $V_c N_c/V_f N_f$) increases absorption in the coolant and thus decreases k_∞; the presence of structural material has the same effect. Equally important, since coolant and structural materials have lower atomic weights than the fuel, neutron collisions with these nuclei degrade the neutron energy. Thus the more coolant is present, the more degraded the energy spectrum will become. The degraded spectrum impacts Eq. (4.16) primarily though the energy-averaged values of the fuel cross sections. For as Fig. 3.1 indicates, as the energy of the neutrons decreases, so does the ratio of $\nu\sigma_f$ to σ_a, thus decreasing the multiplication.

4.4   Thermal Reactor Lattices

Equation 4.1 serves as a starting point for the treatment of thermal as well as fast reactor lattices. However, in the thermal and intermediate neutron energy ranges, which are central to an understanding of thermal reactor physics, cross sections generally are larger than for higher energy neutrons of primary interest in fast reactor physics. In addition the dimensions of coolant and moderator regions typically are larger than those of the coolant channels in fast reactors. The net result of these two factors is that the diameters of fuel pins and the lateral dimensions of moderator and/or coolant may measure several mean free paths or more. In such circumstances the magnitudes of the flux in fuel and moderator regions may differ significantly, with the flux becoming depressed in the fuel region over energy ranges where the fuel absorption cross section is large.

For clarity, we consider a simple two volume model, where V_f and V_m are the fuel and moderator volumes, and hence $V = V_f + V_m$. In doing this we have assumed that the moderator is also the coolant and occupies the same volume, V_m. The model may be generalized to treat separate regions of coolant and moderator, as well as to account for smaller amounts of structural and control materials.

Our simplified model accounts for the difference in fluxes in fuel and moderator by dividing the cellular reaction rates into fuel and moderator contributions:

$$V\Sigma_x(E)\varphi(E) = V_f\Sigma_x^f(E)\varphi_f(E) + V_m\Sigma_x^m(E)\varphi_m(E), \qquad (4.17)$$

where x denotes the reaction type, and $\varphi_f(E)$, $\varphi_m(E)$, and $\varphi(E)$ are the spatial averages over V_f, V_m, and V, respectively. Substituting Eq. (4.17) into Eq. (4.1) yields

$$k_\infty = \frac{V_f \int_0^\infty \nu\Sigma_f^f(E)\varphi_f(E)dE}{V_f \int_0^\infty \Sigma_a^f(E)\varphi_f(E)dE + V_m \int_0^\infty \Sigma_a^m(E)\varphi_m(E)dE}, \qquad (4.18)$$

where only a single term appears in the numerator since $\Sigma_f^m(E) = 0$.

Further analysis is facilitated by dividing the energy spectrum into the thermal (T), intermediate (I), and fast (F) ranges introduced in Chapter 3. As before, we take $1.0\,\text{eV}$ and $0.1\,\text{MeV}$ as convenient partition points between thermal, intermediate, and fast neutrons.

Reaction rates then divide into these three ranges, with each having a distinct pattern of fission and capture:

$$\int_0^\infty \Sigma_x^y(E)\varphi_y(E)dE = \int_T \Sigma_x^y(E)\varphi_y(E)dE + \int_I \Sigma_x^y(E)\varphi_y(E)dE$$
$$+ \int_F \Sigma_x^y(E)\varphi_y(E)dE. \tag{4.19}$$

Fission takes place primarily in the thermal neutron range, with a smaller amount added from fast fission in the fertile material. Thus we delete the intermediate range from fission reactions and write

$$\int_0^\infty \nu\Sigma_f^f(E)\varphi_f(E)dE \approx \int_T \nu\Sigma_f^f(E)\varphi_f(E)dE + \int_F \nu\Sigma_f^f(E)\varphi_f(E)dE. \tag{4.20}$$

Since moderator materials have significant absorption cross sections only for thermal neutrons we make the further simplification

$$\int_0^\infty \Sigma_a^m(E)\varphi_m(E)dE \approx \int_T \Sigma_a^m(E)\varphi_m(E)dE. \tag{4.21}$$

Finally, fuel absorbs both intermediate neutrons—through resonance capture—and thermal neutrons, whereas fast neutron absorption is minimal. Thus

$$\int_0^\infty \Sigma_a^f(E)\varphi_f(E)dE \approx \int_T \Sigma_a^f(E)\varphi_f(E)dE + \int_I \Sigma_a^f(E)\varphi_f(E)dE. \tag{4.22}$$

These simplifications reduce Eq. (4.18) to the more explicit form:

$$k_\infty = \frac{V_f\left[\int_T \nu\Sigma_f^f(E)\varphi_f(E)dE + \int_F \nu\Sigma_f^f(E)\varphi_f(E)dE\right]}{V_f\left[\int_T \Sigma_a^f(E)\varphi_f(E)dE + \int_I \Sigma_a^f(E)\varphi_f(E)dE\right] + V_m\int_T \Sigma_a^m(E)\varphi_m(E)dE}. \tag{4.23}$$

The Four Factor Formula

Although Eq. (4.23) brings out the importance of thermal neutrons (three of the five integrations are over the thermal range), the central role of the neutron slowing down in determining the lattice multiplication is contained within it only implicitly. To make the physical processes more explicit a simplified model—the four factor formula for k_∞—was

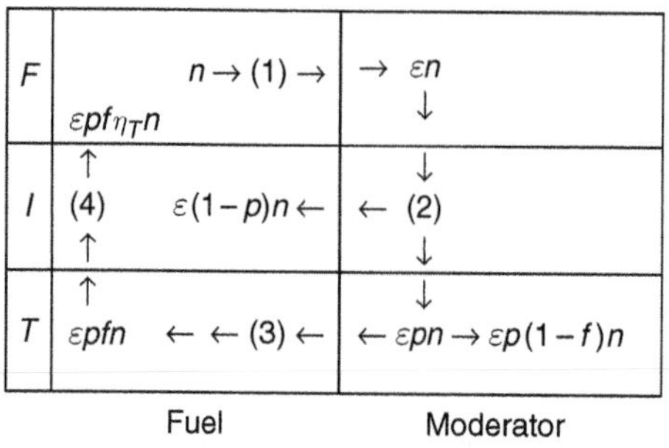

F $\varepsilon pf\eta_T n$	$n \to (1) \to$	$\to \varepsilon n$ $\downarrow$
I (4) $\uparrow$ $\uparrow$	$\uparrow$ $\varepsilon(1-p)n \leftarrow$	$\downarrow$ $\leftarrow (2)$ $\downarrow$
T εpfn	$\uparrow$ $\leftarrow \leftarrow (3) \leftarrow$	$\downarrow$ $\leftarrow \varepsilon pn \to \varepsilon p(1-f)n$

Fuel   Moderator

FIGURE 4.5   Four factor formula for a thermal reactor neutron cycle.

developed early in the history of reactor physics. Based on physical arguments and related to measurements that could be performed at the time, the four factor formula remains valuable tool in understanding the neutron cycle in thermal reactors, and particularly in relating neutron behavior to the thermal hydraulic feedback discussed in Chapter 9. In what follows we present the four factors first qualitatively and then quantitatively. Then, in the final subsection, we employ the formula to examine enrichment, moderator to fuel volume ratio, and other design parameters in determining the multiplication of a pressurized water reactor lattice.

Figure 4.5 illustrates schematically the behavior of neutrons in a thermal reactor lattice consisting of fuel and moderator. The horizontal axis denotes the radial distance from the center to the outside of a lattice cell consisting of a cylindrical fuel element surrounded by moderator; it is separated into fuel and moderator regions. The diagram's vertical axis demarks the neutron energy ranging from 0.01 eV to 10 MeV, with the neutron energies divided into thermal (T), intermediate (I), and fast (F) ranges.

Most of the fission neutrons are born as a result of the absorption of thermal neutrons in the fuel, and they emerge as fast neutrons. Assume that n such fast neutrons originate in the fuel as indicated in Fig. 4.5. Some nominal fraction of these neutrons will cause fast fission in the fertile material, resulting in a total number of εn ($\varepsilon > 1$) fast neutrons produced from fission, where ε is the fast fission factor. The εn fission neutrons migrate into the moderator region as step 1 of the diagram indicates. They then undergo slowing down, indicated as step 2, as a result of collisions with the light atomic weight moderator nuclei. However only some fraction p survive to thermal energies, with the remaining neutrons lost to the resonance capture in the fuel; p (<1) is referred to as the resonance escape probability. Of the εn neutrons which undergo slowing down, εpn arrive at thermal energies, while $\varepsilon(1-p)n$ are lost to capture in the resonances. After arrival at thermal energy some of the neutrons

TABLE 4.2
Representative Four Factor and k_∞ Values for Thermal Reactors

	PWR Pressurized-H$_2$O Reactor	*BWR Boiling-H$_2$O Reactor*	*PHWR CANDU-D$_2$O Reactor*	*HTGR C-Moderated Reactor*
ε	1.27	1.28	1.08	1.20
p	0.63	0.63	0.84	0.62
f	0.94	0.94	0.97	0.98
η_T	1.89	1.89	1.31	2.02
k_∞ [a]	1.41	1.40	1.12	1.47

[a] Fresh fuel without neutron poisons.
Source: Data courtesy of W. S. Yang, Argonne National Laboratory.

are absorbed by the moderator and are thereby lost. However in step 3 a larger fraction f (<1), which is referred to as the thermal utilization, enters the fuel and is absorbed. Thus the fuel absorbs εpfn while the moderator absorbs $\varepsilon p (1 - f)n$ neutrons. For each thermal neutron absorbed in the fuel, η_T (>1) fission neutrons result. Thus, as step 4 symbolizes, $\varepsilon pf\eta_T n$ fission neutrons appear from thermal fission with MeV energies, generated from the n such fission neutrons of the previous generation. Thus according to the definition of Eqs. (3.1) and (3.2):

$$k_\infty = \varepsilon pf\eta_T. \qquad (4.24)$$

Table 4.2 displays representative values of the four factors contributing to k_∞ for the major classes of thermal reactors. The values of k_∞, which are substantially greater than one, represent normal operating conditions. But understand that they are calculated for fresh fuel in the absence of all control poisons. In an operating reactor, of course, fuel depletion, which reduces η_T, the presence of control rods or other control poisons, which reduce f, when taken together with P_{NL}, the nonleakage probability, must yield $k = k_\infty P_{NL} = 1$. To better understand the importance of the neutron moderators we examine each of the four factors more quantitatively in terms of the neutron flux and cross sections in the thermal (T), intermediate (I), and fast (F) energy ranges.

Fast Fission Factor

The fast fission factor is the ratio of total fission neutrons produced to the thermal fission neutrons. Since the neutrons produced by fission of intermediate energy neutrons can be neglected, we have

$$\varepsilon = \frac{\int_T \nu\Sigma_f^f(E)\varphi_f(E)dE + \int_F \nu\Sigma_f^f(E)\varphi_f(E)dE}{\int_T \nu\Sigma_f^f(E)\varphi_f(E)dE}, \qquad (4.25)$$

or equivalently,

$$\varepsilon = 1 + \frac{\int_F \nu\Sigma_f^f(E)\varphi_f(E)dE}{\int_T \nu\Sigma_f^f(E)\varphi_f(E)dE}, \qquad (4.26)$$

where the ratio of integrals on the right varies significantly with the moderator and fuel enrichment employed, ranging between 0.02 and 0.30.

Resonance Escape Probability

All of the fast neutrons scattered downward in energy are absorbed either in the intermediate energy range by the fuel's resonance capture cross sections or in the thermal energy range by fuel or moderator. Because neutrons captured by the fuel resonances are lost from the slowing down process, the fraction that survives to thermal energies is the resonance escape probability:

$$p = \frac{V_f \int_T \Sigma_a^f(E)\varphi_f(E)dE + V_m \int_T \Sigma_a^m(E)\varphi_m(E)dE}{V_f \left[\int_T \Sigma_a^f(E)\varphi_f(E)dE + \int_I \Sigma_a^f(E)\varphi_f(E)dE \right] + V_m \int_T \Sigma_a^m(E)\varphi_m(E)dE}. \qquad (4.27)$$

Adding and subtracting $V_f \int_I \Sigma_a^f(E)\varphi_f(E)dE$ to the numerator, we may rewrite the escape probability as

$$p = 1 - \frac{V_f \int_I \Sigma_a^f(E)\varphi_f(E)dE}{V_f \left[\int_T \Sigma_a^f(E)\varphi_f(E)dE + \int_I \Sigma_a^f(E)\varphi_f(E)dE \right] + V_m \int_T \Sigma_a^m(E)\varphi_m(E)dE}. \qquad (4.28)$$

The total absorption, which is the denominator, must equal the total number of neutrons slowing down, or Vq, where q is the slowing down density—introduced in Chapter 3—averaged over the cell. Next, divide the cell-averaged slowing down density into contributions q_f and q_m from fuel and moderator regions:

$$q = \frac{V_f}{V}q_f + \frac{V_m}{V}q_m. \tag{4.29}$$

Because the fuel nuclei have much larger atomic weights than those of the moderator, slowing down in the fuel can be neglected to first approximation. We then have

$$Vq \approx V_m q_m. \tag{4.30}$$

Replacing the denominator in Eq. (4.28) by $V_m q_m$ then yields

$$p = 1 - \frac{V_f}{V_m q_m}\int_I \Sigma_a^{fe}(E)\varphi_f(E)\,dE, \tag{4.31}$$

where we have also replaced $\Sigma_a^f(E)$ with $\Sigma_a^{fe}(E)$ since the dominant resonance capture takes place in the fertile material.

In the intermediate energy range moderators may be approximated as purely scattering materials. In these circumstances Eq. (3.26) relates the flux and slowing down density; if the scattering cross section Σ_s^m of the moderator is energy independent then the flux is $1/E$, and Eq. (3.26) reduces to

$$q_m = \xi^m \Sigma_s^m E\varphi_m(E), \tag{4.32}$$

where ξ^m is the moderator slowing down decrement defined in Chapter 2. Inserting this expression into Eq. (4.31) then yields

$$p = 1 - \frac{V_f}{V_m \xi^m \Sigma_s^m E\varphi_m(E)}\int_I \Sigma_a^{fe}(E')\varphi_f(E')\,dE'. \tag{4.33}$$

Because q_m and $\xi^m \Sigma_s^m$ are constants, Eq. (4.32) indicates that $E\varphi_m(E)$ also must be independent of energy. Hence we may move it inside the integral. Then, writing $\Sigma_a^{fe}(E) = N_{fe}\sigma_a^{fe}(E)$, we obtain the customary form of the resonance escape probability:

$$p = 1 - \frac{V_f N_{fe}}{V_m \xi^m \Sigma_s^m}I, \tag{4.34}$$

where the resonance integral is defined by

$$I = \int_I \frac{\sigma_a^{fe}(E)\varphi_f(E)}{E\varphi_m(E)}\,dE. \tag{4.35}$$

Most resonances are sufficiently widely spaced that a $1/E$ flux distribution is reestablished at energies between them, but with a slowing down density decreased in proportion to the fraction of neutrons absorbed. This being the case we may apply the two preceding equations to the ith resonance,

$$p_i = 1 - \frac{V_f N_{fe}}{V_m \zeta^m \Sigma_s^m} I_i, \tag{4.36}$$

and

$$I_i = \int_I \frac{\sigma_{ai}^{fe}(E) \varphi_f(E)}{E \varphi^m(E)} dE. \tag{4.37}$$

and write the resonance escape probability as the product of the p_i. If T is the total number of resonances, then

$$p = p_1 p_2 p_3 \cdots p_i \cdots p_{T-1} p_T. \tag{4.38}$$

Generally, the escape probability for a single resonance is sufficiently close to one that Eq. (4.36) is a reasonable two-term approximation to the exponential function. Thus we take

$$p_i = \exp\left(-\frac{V_f N_{fe}}{V_m \zeta^m \Sigma_s^m} I_i\right), \tag{4.39}$$

and inserting this result into Eq. (4.38) gives

$$p = \exp\left(-\frac{V_f N_{fe}}{V_m \zeta^m \Sigma_s^m} I\right), \tag{4.40}$$

where the resonance integral is the sum of contributions from the individual resonances:

$$I = \sum_{i=1}^{T} I_i. \tag{4.41}$$

Figure 3.4 illustrated that self-shielding in energy reduces the neutron capture in the resonances of the fertile material. In fact, the separation of fuel from moderator increases this desirable effect with space–energy self-shielding which depresses the ratio $\varphi_f(E)/\varphi_m(E)$ in Eq. (4.37) at energies were the resonance peaks appear in the absorption cross section $\sigma_a^{fe}(E)$. Figure 4.6 illustrates spatial self-shielding as well as the effect of temperature on it. A resonance cross section and the Doppler broadening that takes place as the temperature

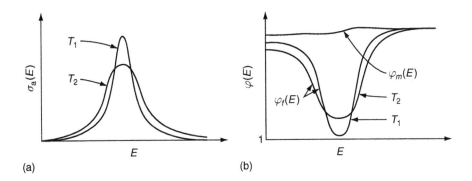

FIGURE 4.6   Effect of Doppler broadening on self-shielding for $T_2 > T_1$. (a) Resonance absorption cross section, (b) neutron flux in fuel and moderator.

increases from T_1 to T_2 is illustrated in Fig. 4.6a. We see in Fig. 4.6b the depression in $\varphi_f(E)$, the flux in the fuel, relative to that of $\varphi_m(E)$, the flux in the moderator. As indicated, increasing the temperature decreases the self-shielding at the resonance peak, and this in turn increases the absorption as the temperature increases.

Only the increase in the resonance escape probability caused by the separations of fuel from moderator allowed reactors fueled with natural uranium and graphite moderators to be built. Without the spatial self-shielding provided by the separation of fuel and moderator, values of $k_\infty \geq 1$ are possible with natural uranium fuel only if heavy water is the moderator. Even with spatial self-shielding, slightly enriched uranium is required to achieve criticality in a light water–moderated system.

More advanced texts include detailed analysis of space–energy self-shielding. However, experimental investigations have led to empirical equations for the resonance integrals of uranium-238 fuel rods given in Table 4.3. The resonance integrals are in barns, the density ρ in g/cm³, and D is the rod diameter in cm. These equations are valid for isolated rods with $0.2\,\text{cm} < D < 3.5\,\text{cm}$; if the rods are located in a tightly packed lattice the self-shielding increases somewhat through what is called a Dancoff correction. In either case, the

TABLE 4.3
Resonance Integrals for Fuel Rods

$I = 2.95 + 25.8\sqrt{4/\rho D}$	for U metal
$I = 4.45 + 26.6\sqrt{4/\rho D}$	for UO_2

Source: Nordheim, L. W., "The Doppler Coefficient," *The Technology of Nuclear Reactor Safety*, T. J. Thompson and J. G. Beckerley, eds., Vol. 1, MIT Press, Cambridge, MA, 1966.

spatial self-shielding is evident, since as the fuel diameter increases, the resonance integral—and therefore the absorption—decreases.

Thermal Utilization and η_T

The thermal utilization, f, is the fraction of thermal neutrons absorbed in the fuel. Since all the thermal neutrons must be absorbed in either fuel or moderator, we have

$$f = \frac{V_f \int_T \Sigma_a^f(E)\varphi_f(E)dE}{V_f \int_T \Sigma_a^f(E)\varphi_f(E)dE + V_m \int_T \Sigma_a^m(E)\varphi_m(E)dE}. \tag{4.42}$$

Finally, the number of fission neutrons produced per thermal neutron absorbed in the fuel is

$$\eta_T = \frac{\int_T \nu\Sigma_f^f(E)\varphi_f(E)dE}{\int_T \Sigma_a^f(E)\varphi_f(E)dE}. \tag{4.43}$$

The expressions for f and η_T simplify considerably by defining cross sections averaged only over the thermal neutron spectrum. Let

$$\bar{\varphi}_{fT} = \int_T \varphi_f(E)dE \tag{4.44}$$

and

$$\bar{\varphi}_{mT} = \int_T \varphi_m(E)dE \tag{4.45}$$

be the thermal fluxes, space-averaged, respectively, over the fuel and moderator regions. The thermal cross sections for fuel and moderator then become

$$\bar{\Sigma}_{xT}^f = \bar{\varphi}_{fT}^{-1} \int_T \Sigma_x^f(E)\varphi_f(E)dE \tag{4.46}$$

and

$$\bar{\Sigma}_{xT}^m = \bar{\varphi}_{mT}^{-1} \int_T \Sigma_x^m(E)\varphi_m(E)dE. \tag{4.47}$$

With these thermal cross section definitions f and η_T simplify to

$$f = \frac{1}{1 + \varsigma\left(V_m \bar{\Sigma}_{aT}^m \middle/ V_f \bar{\Sigma}_{aT}^f\right)} \qquad (4.48)$$

and

$$\eta_T = \frac{\nu \bar{\Sigma}_{fT}^f}{\bar{\Sigma}_{aT}^f}, \qquad (4.49)$$

where the thermal disadvantage factor is defined as the ratio of thermal neutron flux in the moderator to that in the fuel:

$$\varsigma = \bar{\varphi}_{mT} / \bar{\varphi}_{fT}. \qquad (4.50)$$

The disadvantage is that the more neutrons that are captured in the moderator because of the larger flux there, the fewer will be available to create fission in the fuel.

k_∞ Reconsidered

The question remains, How does the four factor formula relate to the value of k_∞ given by Eq. (4.23)? For an answer, insert Eqs. (4.25), (4.27), (4.42), and (4.43) for ε, p, f, and η_T into $k_\infty = \varepsilon p f \eta_T$. Canceling terms, we see that the result is identical to Eq. (4.23):

$$k_\infty = \frac{\int_T \nu\Sigma_f^f(E)\varphi_f(E)dE + \int_F \nu\Sigma_f^f(E)\varphi_f(E)dE}{\cancel{\int_T \nu\Sigma_f^f(E)\varphi_f(E)dE}}$$

$$\bullet \, \frac{\cancel{V_f\int_T \Sigma_a^f(E)\varphi_f(E)dE + V_m\int_T \Sigma_a^m(E)\varphi_m(E)dE}}{V_f\left[\int_I \Sigma_a^f(E)\varphi_f(E)dE + \int_T \Sigma_a^f(E)\varphi_f(E)dE\right] + V_m\int_T \Sigma_a^m(E)\varphi_m(E)dE}$$

$$\bullet \, \frac{V_f\int_T \Sigma_a^f(E)\varphi_f(E)dE}{\cancel{V_f\int_T \Sigma_a^f(E)\varphi_f(E)dE + V_m\int_T \Sigma_a^m(E)\varphi_m(E)dE}} \bullet \frac{\cancel{\int_T \nu\Sigma_f^f(E)\varphi_f(E)dE}}{\int_T \Sigma_a^f(E)\varphi_f(E)dE}.$$

$$\qquad (4.51)$$

Thus Eq. (4.23) is consistent with the four factor formula.

Pressurized Water Reactor Example

Two of the primary determinants of the thermal reactor core multiplication are the fuel enrichment and the volume ratio of moderator to fuel. The multiplication increases monotonically with enrichment but exhibits a maximum when plotted with respect to the moderator to fuel ratio. Under- and overmoderated lattices refer to those in which the moderator to fuel ratios are less than or greater than the optimal value. We illustrate the effects of enrichment, moderator to fuel ratio, and other design parameters on multiplication using UO_2 pressurized water reactor lattices as examples.

We begin by expressing the four factors in terms of the enrichment and the moderator to fuel atom ratio. Equations (4.9) and (4.10) define the enrichment $\tilde{e}$ in terms of the number fissile (fi) and fertile (fe) atom densities. Thus analogous to Eq. (4.14) for fast reactors, for thermal reactors we have

$$\bar{\sigma}_{aT}^{f} = \tilde{e}\bar{\sigma}_{aT}^{fi} + (1 - \tilde{e})\bar{\sigma}_{aT}^{fe}. \tag{4.52}$$

Now, however, the energy average is taken over only the thermal range as indicated by the subscript T. With this nomenclature, we may express η_T in terms of the enrichment and η_{fi} of the fissile material as

$$\eta_T = \eta_T^{fi} \Big/ \left[1 + (1 - \tilde{e})\bar{\sigma}_{aT}^{fe}/\tilde{e}\bar{\sigma}_{aT}^{fi}\right]. \tag{4.53}$$

The resonance escape probability is a function of both the enrichment and the ratio of moderator to fuel nuclei, $V_m N_m / V_f N_f$. Because $N_{fe} = (1 - \tilde{e})N_f$, we may rewrite Eq. (4.40) as

$$p = \exp\left(-\frac{(1 - \tilde{e})}{(V_m N_m / V_f N_f)} \frac{I}{\xi \sigma_s^m}\right), \tag{4.54}$$

where $\Sigma_s^m = N_m \sigma_s^m$. Expressed in terms of the moderator to fuel atom ratio the thermal utilization given by Eq. (4.48) becomes

$$f = \frac{1}{1 + \varsigma(V_m N_m / V_f N_f)(\bar{\sigma}_{aT}^m / \bar{\sigma}_{aT}^f)}, \tag{4.55}$$

where we have again expressed macroscopic cross section in terms of nuclide densities and microscopic cross sections: $\bar{\Sigma}_{aT}^m = N_m \bar{\sigma}_{aT}^m$

and $\bar{\Sigma}_{aT}^f = N_f \bar{\sigma}_{aT}^f$. Finally the fast fission factor, given by Eq. (4.26), appears as

$$\varepsilon = 1 + \frac{(1 - \tilde{e})}{\tilde{e}} \frac{\nu^{fe} \bar{\sigma}_{fF}^{fe}}{\nu^{fi} \bar{\sigma}_{fT}^{fi}} \qquad (4.56)$$

when written in terms of microscopic cross sections.

Two phenomena compete as $V_m N_m / V_f N_f$, the ratio of moderator to fuel, is increased. Equation (4.54) indicates that the resonance escape probability increases. This happens as a result of the greater ability of the moderator to slow down neutrons past the capture resonances. In contrast, Eq. (4.55) shows that a larger value of $V_m N_m / V_f N_f$ results in more thermal neutron capture in the moderator, thus decreasing the thermal utilization. As a result of these opposing phenomena, there exists an optimum moderator to fuel ratio, which for a given enrichment, fuel element size, and soluble absorber concentration yields the maximum value of k_∞. Figure 4.7 illustrates these effects.

The figure also illuminates a number of other effects. Increasing the concentration of boron poison in the moderator reduces the

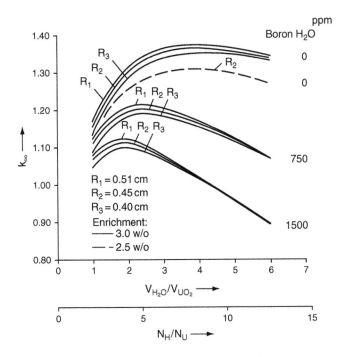

FIGURE 4.7 k_∞ for pressurized water lattices vs moderator to fuel volume ratios with different rod radii, boron concentrations, and enrichments (adapted from *Theorie der Kernreaktoren*, 1982, by D. Emendorfer and K. H. Hocker, by permission of VGB, Essen).

multiplication, and the effect becomes larger—as one would expect—as the moderator to fuel volume ratio increases. Likewise, increasing the fuel rod radius increases the multiplication, but by a smaller amount. This comes about from the increase in spatial self-shielding of the capture resonances: Increased self-shielding reduces the resonance integral given in Table 4.3, which subsequently increases the resonance escape probability displayed in Eq. (4.40). Finally, enrichment has a marked effect in increasing k_∞, which is apparent from observing the curves for 2.5% (dashed) and 3.0% (solid) enrichment with the other parameters held constant.

In liquid-moderated systems the core invariably is designed to be undermoderated—to have less moderator than the optimum—under operating conditions. This is to ensure stability. Because a liquid expands more rapidly than the solid fuel as the core temperature increases, the ratio of moderator to fuel atoms will decrease with temperature. This will move the operating points on the curves in Fig. 4.7 to the left. Thus the value of k_∞ will decrease with increasing temperature in an undermoderated core, creating the negative feedback required for a stable system. Conversely an overmoderated core would create a positive temperature feedback, and result in an unstable system, unless a negative feedback mechanism, such as the Doppler broadening discussed in Chapter 3, overrides the positive effect. As we will elaborate in Chapter 9, however, too much negative feedback can create other challenges. Reactor designers must balance the composite effects of moderator density, fuel temperature, and other phenomena to ensure system stability under all operation conditions.

The situation may be more complex in solid-moderated reactors, such as those employing graphite, because it is the relative values of the expansion coefficients of two solids that then come into play. Moreover, although the atom density of gas coolant may be too small to have a measurable impact, the relative expansion of the coolant channel diameter may become a significant factor. If a liquid coolant is used in combination with graphite, such as in the RBMK reactors, then the interactions are more complex yet, for the interacting temperature effects on fuel, coolant, and moderator must be considered separately, and in combination.

Bibliography

Bell, George I., and Samuel Glasstone, *Nuclear Reactor Theory*, Van Nostrand Reinhold, NY, 1970.

Bonalumi, R. A., "In-Core Fuel Management in CANDU-PHR Reactors," *Handbook of Nuclear Reactors Calculations, II*, Yigan Ronen, ed., CRC Press, Boca Raton, FL, 1986.

Duderstadt, James J., and Louis J. Hamilton, *Nuclear Reactor Analysis*, Wiley, NY, 1976.

Emendorfer, D., and K. H. Hocker, *Theorie der Kernreaktoren*, Bibliographisches Institut AG, Zurich, 1982 .

Glasstone, Samuel, and Alexander Sesonske, *Nuclear Reactor Engineering*, 3rd ed., Van Nostrand Reinhold, NY, 1981.

Lamarsh, John, and Anthony J. Baratta, *Introduction to Nuclear Engineering*, 3rd ed., Prentice-Hall, 2001.

Nordheim, L. W., "The Doppler Coefficient," *The Technology of Nuclear Reactor Safety*, T. J. Thompson and J. G. Beckerley, eds., Vol. 1, MIT Press, Cambridge, MA, 1966.

Salvatores, Max, "Fast Reactor Calculations," *Handbook of Nuclear Reactors Calculations, III*, Yigan Ronen, ed., CRC Press, Boca Raton, FL, 1986.

Steward, H. B., and M. H. Merril, "Kinetics of Solid-Moderators Reactor", *Technology of Nuclear Reactor Safety*, T. J. Thompson and J. G. Beckerley, eds., Vol. 1, MIT Press, Cambridge, MA, 1966.

Turinsky, Paul J., "Thermal Reactor Calculations," *Handbook of Nuclear Reactors Calculations, III*, Yigan Ronen, ed., CRC Press, Boca Raton, FL, 1986.

Problems

4.1. A reactor is to be built with fuel rods of 1.2 cm in diameter and a liquid moderator with a 2:1 volume ratio of moderator to fuel. What will the distance between nearest fuel centerlines be

   a. For a square lattice?
   b. For a hexagonal lattice?

4.2. In a fast reactor designers often want to minimize the coolant to fuel volume ratio to minimize the amount of neutron slowing down. From a geometric point of view what is the theoretical limit on the smallest ratio of coolant to fuel volume that can be obtained

   a. With a square lattice?
   b. With a hexagonal lattice?

4.3. A sodium-cooled fast reactor is fueled with PuO_2, mixed with depleted UO_2. The structural material is iron. Averaged over the spectrum of fast neutrons, the microscopic cross sections and densities are as follows:

	σ_f b	σ_a b	σ_t b	ρ g/cm^3
PuO_2	1.95	2.40	8.6	11.0
UO_2	0.05	0.404	8.2	11.0
Na	–	0.0018	3.7	0.97
Fe	–	0.0087	3.6	7.87

The fuel is 15% PuO_2 and 85% UO_2 by volume. The volumetric composition of the core is 30% fuel, 50% coolant, and 20% structural material. Calculate k_∞ assuming that the values of ν for plutonium and uranium in the fast spectrum are 2.98 and 2.47, respectively, and that the cross sections of oxygen can be neglected. What fraction of the mass of the core does the fuel account for?

4.4. Suppose the nonleakage probability for a sodium-cooled fast reactor specified in problem 4.3 is 0.90. Using the data from problem 4.3, adjust the volume fractions of PuO_2 and UO_2 in the fuel so that $k = 1.0$. What is the % PuO_2 in the fuel by volume?

4.5. Verify Eq. (4.18).

4.6. A pressurized water reactor has 3% enriched UO_2 fuel pins that are 1.0 cm in diameter and have a density of $11.0\,g/cm^3$. The moderator to fuel volume ratio is 2:1. Calculate η_T, p, f, and k_∞ at room temperature under the assumptions that $\varepsilon = 1.24$, the thermal disadvantage factor $\varsigma = 1.16$, and the Dancoff correction increases the fuel diameter for the resonance integral calculation by 10%. (Use fuel data from Problem [3.4].)

4.7. Suppose the fuel rods from problem 4.6 are to be used in a D_2O-moderated reactor.

a. What volume ratio of moderator to fuel is required to give the same value of p as for the H_2O lattice in problem 4.6? (Assume no Dancoff correction.)

b. What volume ratio of moderator to fuel is required to give the same value of f as for the H_2O lattice in problem 4.6? (Assume ς is unchanged.)

4.8. A reactor lattice consists of uranium rods in a heavy water moderator. The heavy water is replaced by light water.

a. Would the resonance escape probability increase or decrease? Why?

b. Would the thermal utilization increase or decrease? Why?

c. What would you expect the net effect on k_∞ to be? Why?

4.9. Suppose the volume ratio of coolant to fuel is increased in a pressurized water reactor:

a. Will the fast fission factor increase, decrease, or remain unchanged? Why?

b. Will the resonance escape probability increase, decrease, or remain unchanged? Why?

    c. Will the thermal utilization increase, decrease, or remain unchanged? Why?

    d. Will the value of η_T increase, decrease, or remain unchanged? Why?

4.10. Using the data from problem 4.6, vary the coolant to fuel volume ratio between 0.5 and 2.5 and plot the following vs V_m/V_f:

    a. The resonance escape probability.

    b. The thermal utilization.

    c. k_∞.

    d. Determine the moderator to fuel volume ratio that yields the largest k_∞.

    e. What is the largest value of k_∞?

You may assume that changes in the fast fission factor and the thermal disadvantage factor are negligible.

4.11. A reactor designer decides to replace uranium with UO_2 fuel in a water-cooled reactor, keeping the enrichment, fuel diameter, and water to fuel volume ratios the same.

    a. Will p increase, decrease, or remain unchanged? Why?

    b. Will f increase, decrease, or remain unchanged? Why?

    c. Will η_T increase, decrease, or remain unchanged? Why?

4.12. The fuel for a thermal reactor has the following composition by atom ratio: 2% uranium-235, 1% plutonium-239, and 97% uranium-238. Calculate the value of η_T to be used for this fuel in the four factor formula. (Use the data given for problem 3.4.)

CHAPTER 5

Reactor Kinetics

5.1 Introduction

This chapter takes up a detailed examination of the time-dependent behavior of neutron chain reactions. In order to emphasize the time variable we make two simplifications that remove the need to treat concomitantly the neutron energy and spatial variables. First, we assume that the techniques of Chapters 3 and 4 have been applied to average both the neutron distribution and associated cross sections over energy. Second, we delay the explicit treatment of spatial effects until the following chapters, assuming for now that neutron leakage from the system is either negligible or can be treated by the nonescape probability approximation introduced earlier.

We begin by introducing a series of neutron balance equations and their time-dependent behavior. First we examine a system in which no fissionable materials are present; such a system is designated as nonmultiplying. Subsequently, we include fissionable isotopes, and examine the behavior of the resulting multiplying system. In both cases we assume that neutron leakage from the systems can be ignored. We then incorporate the effects of leakage, in order to examine the phenomena of criticality in systems of finite size.

The most noteworthy simplification in these equations is the assumption that all neutrons are produced instantaneously at the time of fission. In reality a small fraction of fission neutrons are delayed because they are emitted as a result of the decay of certain fission products. These delayed neutrons have profound effects on the behavior of chain reactions. The remainder of the chapter deals with the reactor kinetics that results from the combined effects of fission neutrons, both those produced promptly and those that are delayed.

5.2   Neutron Balance Equations

To obtain neutron balance equations we utilize the following definitions:

$n(t) = $ total number of neutrons at time t,
$\bar{v} = $ average neutron speed,
$\Sigma_x = $ energy-averaged cross section for reaction type x.

Here $n(t)$ refers to all of the neutrons in the system, regardless of their location or kinetic energy. Likewise $\bar{v}$ and Σ_x are the results obtained from averaging over all neutron energies, and in the case of Σ_x we also assume that in a power reactor it has been spatially averaged over the lattice cell.

Infinite Medium Nonmultiplying Systems

We first determine the time rate of change of the number of neutrons, $n(t)$, in a nonmultiplying system, that is, a system containing no fissionable material. In addition, we assume that the system's dimensions are so large that the small fraction of neutrons that leak from its surface can be neglected. The neutron balance equation is then

$$\frac{d}{dt}n(t) = \text{\# of source neutrons produced/s} - \text{\# of neutrons absorbed/s.} \tag{5.1}$$

We replace the first term on the right by the source $S(t)$, defined as the number of neutrons introduced per second. Recall that the macroscopic absorption cross section Σ_a is the probability that a neutron will be absorbed per cm of travel. Thus $\bar{v}\Sigma_a$ is the probability per second that a neutron will be absorbed, and $\Sigma_a \bar{v}n(t)$ is the number of neutrons absorbed per second. Equation (5.1) becomes

$$\frac{d}{dt}n(t) = S(t) - \Sigma_a \bar{v}n(t). \tag{5.2}$$

To determine the average neutron lifetime between birth and absorption we assume that at $t = 0$ the system contains $n(0)$ neutrons, but no further neutrons are produced. Thus $S(t) = 0$ in Eq. (5.2):

$$\frac{d}{dt}n(t) = -\Sigma_a \bar{v}n(t). \tag{5.3}$$

Analogous in form to the radioactive decay equation in Chapter 1, Eq. (5.3) has the solution

$$n(t) = n(0) \exp(-t/l_\infty),$$ (5.4)

where we have defined

$$l_\infty = 1/\bar{v}\Sigma_a.$$ (5.5)

Because l_∞ has the units of time, we designate it as the neutron lifetime. The subscript indicates that we are referring to an infinite medium; the lifetime is not shortened as a result of the neutrons leaking from the system. More formal justification for designating l_∞ as the neutron lifetime derives from the definition of $\bar{t}$, the mean neutron lifetime:

$$\bar{t} \equiv \frac{\displaystyle\int_0^\infty tn(t)dt}{\displaystyle\int_0^\infty n(t)dt} = 1/\bar{v}\Sigma_a = l_\infty.$$ (5.6)

Using this definition for neutron lifetime we may write Eq. (5.2) as

$$\frac{d}{dt}n(t) = S_o - \frac{1}{l_\infty}n(t),$$ (5.7)

where we have assumed the source to be independent of time: $S(t) \to S_o$. Next consider the situation where no neutrons are present until the source is inserted at $t=0$. Solving Eq. (5.7) with the initial condition $n(0) = 0$ yields

$$n(t) = l_\infty S_o[1 - \exp(-t/l_\infty)].$$ (5.8)

Thus the neutron population first increases but then stabilizes at $n(\infty) = l_\infty S_o$.

The foregoing equations illustrate that the speed at which neutron populations build or decay depends strongly on the neutron lifetime. Accordingly, the processes take place very rapidly in non-multiplying systems, because neutron lifetimes in them range from 10^{-8} to 10^{-4} s. The longer lifetimes take place in systems in which the neutrons reach thermal energies before being absorbed, for then the speed in Eq. (5.5) is dominated by the slow neutrons. Systems in which most of the neutrons are absorbed at higher energies, before slowing down can take place, have much shorter lifetimes.

Infinite Medium Multiplying Systems

Next consider a multiplying system, that is, one in which fissionable material is present. We again neglect leakage effects by assuming that the system's dimensions are infinitely large. For now we also assume that all neutrons from fission are emitted instantaneously, thus ignoring the small but important fraction of fission neutrons that undergo delayed emission. With these assumptions the neutron balance equation becomes

$$\frac{d}{dt}n(t) = \text{\# of source neutrons produced/s}$$
$$+ \text{\# of fission neutrons produced/s} \qquad (5.9)$$
$$- \text{\# of neutrons absorbed/s.}$$

The source and absorption terms have the same form as before. To obtain the fission term, first note that the number of fission reactions per second is $\Sigma_f \bar{v}n(t)$. Then with ν denoting the average number of neutrons per fission, the number of fission neutrons produced per second is $\nu\Sigma_f \bar{v}n(t)$. Hence the balance equation becomes

$$\frac{d}{dt}n(t) = S(t) + \nu\Sigma_f \bar{v}n(t) - \Sigma_a \bar{v}n(t). \qquad (5.10)$$

With the infinite medium multiplication defined by Eq. (3.61) as

$$k_\infty = \nu\Sigma_f/\Sigma_a \qquad (5.11)$$

and l_∞ defined by Eq. (5.5), we may rewrite Eq. (5.10) as

$$\frac{d}{dt}n(t) = S(t) + \frac{(k_\infty - 1)}{l_\infty}n(t). \qquad (5.12)$$

The concepts of criticality and multiplication introduced in earlier chapters are closely related. Suppose we consider a system in which there are no sources of neutrons other than from fission. Accordingly, we set $S = 0$ in Eq. (5.12) and obtain

$$\frac{d}{dt}n(t) = \frac{(k_\infty - 1)}{l_\infty}n(t). \qquad (5.13)$$

A system is defined to be critical if there can exist a time-independent population of neutrons in the absence of external sources. Thus for a system to be critical the derivative on the left of this expression must vanish. Consequently, for the system to be critical the multiplication k_∞ must be equal to one. The system is said to be sub- or supercritical, respectively, if in the absence of external sources, the neutron population decreases or increases with time. Equation (5.13) indicates that an infinite system is subcritical, critical, or supercritical according to whether $k_\infty < 1$, $k_\infty = 1$, or $k_\infty > 1$.

Finite Multiplying Systems

We next reformulate the neutron balance equation for the more realistic situation in which the reactor dimensions are finite, and hence neutron leakage from the system must be taken into account. We begin by rewriting Eq. (5.9) with a loss term appended to account for neutron leakage:

$$
\frac{d}{dt}n(t) = \text{\# of source neutrons produced/s}
$$
$$
+ \text{\# of fission neutrons produced/s} \qquad (5.14)
$$
$$
- \text{\# of neutrons absorbed/s}
$$
$$
- \text{\# neutrons leaking from system/s.}
$$

Only the leakage term differs in form from those found in Eq. (5.10). We write it as $\Gamma\Sigma_a \bar{v}n(t)$, which assumes that the number of neutrons leaking from the system is proportional to the number of neutrons absorbed per second, with proportionality constant Γ. With the leakage expressed as $\Gamma\Sigma_a \bar{v}n(t)$, Eq. (5.14) becomes

$$
\frac{d}{dt}n(t) = S(t) + \nu\Sigma_f \bar{v}n(t) - \Sigma_a \bar{v}n(t) - \Gamma\Sigma_a \bar{v}n(t). \qquad (5.15)
$$

We may cast this equation in a more physically meaningful form by expressing Γ in terms of leakage and nonleakage probabilities. We say that neutrons are "born" either as external source neutrons, S, or from fission. They then undergo a number of scattering collisions and ultimately "die" from one of two fates: they are absorbed or they leak from the system. The probability that they will leak from the system is

$$
P_L = \frac{\Gamma\Sigma_a \bar{v}n}{\Sigma_a \bar{v}n + \Gamma\Sigma_a \bar{v}n} = \frac{\Gamma}{1 + \Gamma}. \qquad (5.16)
$$

Thus the nonleakage probability is $1 - P_L$ or

$$
P_{NL} = \frac{1}{1 + \Gamma}. \qquad (5.17)
$$

Thus since we expect the nonleakage probability to grow and approach one as the reactor becomes very large, Γ decreases with increased reactor size.

The nonleakage probability definition assists in writing Eq. (5.15) in a more compact form. We first multiply it by the nonleakage probability and use Eq. (5.17) to eliminate Γ from the equation:

$$P_{NL}\frac{d}{dt}n(t) = P_{NL}S(t) + P_{NL}(\nu\Sigma_f)\,\bar{v}n(t) - \Sigma_a\,\bar{v}n(t). \qquad (5.18)$$

Equations (5.11) and (5.5) allow the last two terms to be combined and written in terms of k_∞ and l_∞:

$$P_{NL}\frac{d}{dt}n(t) = P_{NL}S(t) + \frac{(P_{NL}k_\infty - 1)}{l_\infty}n(t). \qquad (5.19)$$

Finally, we divide this expression by P_{NL} and define the finite medium multiplication and neutron lifetime as

$$k = P_{NL}k_\infty \qquad (5.20)$$

and

$$l = P_{NL}l_\infty. \qquad (5.21)$$

Equation (5.19) then reduces to

$$\frac{d}{dt}n(t) = S(t) + \frac{(k-1)}{l}n(t), \qquad (5.22)$$

which is identical to Eq. (5.12) if the infinite medium subscripts are removed. Both the multiplication and neutron lifetime are smaller for the finite than for the infinite system, because of the neutrons lost to leakage.

5.3 Multiplying Systems Behavior

Although Eq. (5.22) does not incorporate the effects of delayed neutrons, it provides us with a simplified—but qualitatively correct—description of multiplying systems. The definition of criticality once again comes from the form of the equation in which the external source term, $S(t)$, is set equal to zero:

$$\frac{d}{dt}n(t) = \frac{(k-1)}{l}n(t). \qquad (5.23)$$

If there are neutrons in the system at $t=0$, that is, if $n(0) > 0$, then in the absence of an external source,

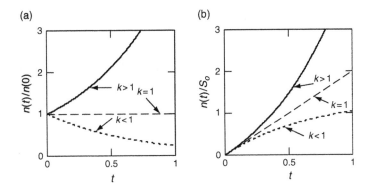

FIGURE 5.1 Neutron populations in subcritical, critical, and supercritical systems. (a) $n(0) = 1$, no source present, (b) $n(0) = 0$, source present.

$$n(t) = n(0) \exp\left[\frac{(k-1)}{l} t\right]. \qquad (5.24)$$

The system is said to be critical if there is a time-independent chain reaction in the absence of sources. Clearly, the system is critical when k, the multiplication, is equal to one. Analogous to the infinite medium, we now have

$$k \begin{cases} >1 & supercritical \\ =1 & critical \\ <1 & subcritical. \end{cases} \qquad (5.25)$$

Figure 5.1a shows the behavior of multiplying systems in the absence of sources.

Considering the case where a time-independent source is present in the system, we must solve Eq. (5.22) with $S(t) \to S_o$. Applying the integrating factor technique explained in Appendix A, with an initial condition $n(0) = 0$, we obtain

$$n(t) = \frac{lS_o}{(k-1)} \left\{ \exp\left[\frac{(k-1)}{l} t\right] - 1 \right\}. \qquad (5.26)$$

Figure 5.1b indicates the behavior of multiplying systems with a time-independent source present. For a supercritical system, $k > 1$, the neutron population rises at an increasing rate, becoming exponential at long times. The behavior of the subcritical system, $k < 1$, becomes more transparent if we first manipulate the minus signs to rewrite Eq. (5.26) in the form

$$n(t) = \frac{lS_o}{(1-k)}\left\{1 - \exp\left[\frac{-(1-k)}{l}t\right]\right\}. \tag{5.27}$$

Thus for $k < 1$ (the subcritical case) the number of neutrons increases at first, but then as the exponential term decays away, the population stabilizes at long times to a time-independent solution of

$$n(\infty) = \frac{lS_o}{(1-k)}. \tag{5.28}$$

The behavior of Eq. (5.26) or Eq. (5.27) is more subtle for the case of a critical system, for which $k = 1$, because then both the term in brackets and the denominator equal zero. However, if we take the limit as $k \to 1$ we can expand the exponential as a power series $\exp(x) = 1 + x + \frac{1}{2}x^2 + \cdots$ in Eq. (5.26) to obtain

$$n(t) = \frac{lS_o}{(k-1)}\left\{1 + \frac{(k-1)}{l}t + \frac{1}{2}\frac{(k-1)^2}{l^2}t^2 + \cdots - 1\right\}. \tag{5.29}$$

Canceling terms, and taking the limit $k \to 1$ yields

$$n(t) = S_0 t. \tag{5.30}$$

Thus the neutron population in a critical system will increase linearly with time in the presence of a time-independent source.

The results of the foregoing considerations are summarized in the Fig. 5.1. There are only two situations in which a time-independent population of neutrons can be established: (1) a critical system in the absence of an external source ($k = 1$ and $S = 0$) and (2) a subcritical system in which there is a source of neutrons present ($k < 1$ and $S > 0$). Note that the definition of criticality depends only on the multiplication k, and not on the source. The behavior of the system, however, depends very strongly on whether a source is present.

Observe that even for very small deviations of the multiplication from one, large time rates of change take place in the neutron population. These are caused by the small value of the neutron lifetime, typically ranging from 10^{-8} to 10^{-4} s, which appears in the denominator of the exponentials. With such small neutron lifetimes, controlling a nuclear reactor would be very difficult if all the neutrons were born instantaneously with fission. Fortunately, the presence of delayed neutrons—which have been neglected until now—greatly decreases the rates of change in the neutron populations to more manageable levels, provided certain restrictions are met.

5.4 Delayed Neutron Kinetics

More than 99% of fission neutrons are produced instantaneously at the time of fission. The remaining fraction, β, results from the decay of the neutron-emitting fission products. These neutron precursors are typically lumped into six groups with half-lives ranging from a fraction of a second to nearly a minute. Table 5.1 displays the six group half-lives and delayed fractions for the three most common fissile isotopes. Note that β is just the sum of the delayed fractions for each group:

$$\beta = \sum_{i=1}^{6} \beta_i. \tag{5.31}$$

If we denote the half-life for the ith group by $t_{i^{1/2}}$, then the average half-life of the delayed neutrons is

$$t_{1/2} = \frac{1}{\beta} \sum_{i=1}^{6} \beta_i t_{i^{1/2}}. \tag{5.32}$$

Moreover, since half-lives and decay constants are related by

$$t_{i^{1/2}} = 0.693/\lambda_i, \tag{5.33}$$

the average decay constant can be defined by

$$\frac{1}{\lambda} = \frac{1}{\beta} \sum_{i=1}^{6} \beta_i \frac{1}{\lambda_i}. \tag{5.34}$$

TABLE 5.1
Delayed Neutron Properties

Approximate Half-life (sec)	Delayed Neutron Fraction		
	U^{233}	U^{235}	Pu^{239}
56	0.00023	0.00021	0.00007
23	0.00078	0.00142	0.00063
6.2	0.00064	0.00128	0.00044
2.3	0.00074	0.00257	0.00069
0.61	0.00014	0.00075	0.00018
0.23	0.00008	0.00027	0.00009
Total delayed fraction	0.00261	0.00650	0.00210
Total neutrons/fission	2.50	2.43	2.90

In the preceding equations the quantity l designates the prompt neutron lifetime, which is the average lifetime of those neutrons produced instantaneously at the time of fission. If we define the average delayed neutron lifetime, l_d, as the time interval between fission and the time at which the delayed neutrons are absorbed or leak from the system, then it has a value of

$$l_d = l + t_{1/2}/0.693 = l + 1/\lambda. \qquad (5.35)$$

Hence including the lifetimes of both prompt and delayed neutrons, we obtain the average neutron lifetime as

$$\bar{l} = (1 - \beta)l + \beta l_d = l + \beta/\lambda. \qquad (5.36)$$

Although delayed neutrons are a small fraction of the total, they dominate the average neutron lifetime, because $\beta/\lambda \gg l$. The delayed neutrons have profound effects on the behavior of chain reactions. But the effects are not adequately described by simply replacing l with $\bar{l}$ in the foregoing equations. Proper treatment requires a set of differential equations to account for the time-dependent behavior of both prompt and delayed neutrons.

Kinetics Equations

To derive a neutron balance equation in which the effects of delayed neutrons are included, we divide the fission terms in Eq. (5.15) into prompt and delayed contributions. If β is the delayed neutron fraction, the rate at which prompt neutrons are produced is $(1 - \beta)\nu\Sigma_f \bar{v}n(t)$. Delayed neutrons, on the other hand, are produced by the decay of fission products. If we define $C_i(t)$ as the number of radioactive precursors producing neutrons with a half-life $t_{i1/2}$, then the rate of delayed neutron production is $\lambda_i C_i(t)$. With the fission term divided into the prompt and delayed contributions, Eq. (5.15) becomes

$$\frac{d}{dt}n(t) = S(t) + (1 - \beta)\nu\Sigma_f \bar{v}n(t) + \sum_i \lambda_i C_i(t) - \Sigma_a \bar{v}n(t) - \Gamma \bar{v}n(t).$$

$$(5.37)$$

We require six additional equations to determine the precursor concentration for each delayed group. Each has the form of a balance equation:

$$\frac{d}{dt}C_i(t) = \#\text{ precursors produced/s} - \#\text{ precursors decaying/s}.$$

$$(5.38)$$

The number of precursors of type i produced per second is $\beta_i \nu \Sigma_f \bar{v} n(t)$, whereas the decay rate is $\lambda_i C_i(t)$. Hence

$$\frac{d}{dt} C_i(t) = \beta_i \nu \Sigma_f \bar{v} n(t) - \lambda_i C_i(t), \quad i = 1, 2, \ldots, 6. \tag{5.39}$$

Taken together, Eqs. (5.37) and (5.39) constitute the neutron kinetics equations. Expressed in terms of the multiplication and prompt neutron lifetime, defined by Eqs. (5.5), (5.11), (5.20), and (5.21), these equations become

$$\frac{d}{dt} n(t) = S(t) + \frac{1}{l}[(1 - \beta)k - 1]n(t) + \sum_i \lambda_i C_i(t) \tag{5.40}$$

and

$$\frac{d}{dt} C_i(t) = \beta_i \frac{k}{l} n(t) - \lambda_i C_i(t), \quad i = 1, 2, \ldots, 6. \tag{5.41}$$

We may ask, under what conditions can these equations have a steady state solution; that is, a solution for which n and C_i are time-independent, causing the derivatives on the left vanish? For a time-independent source, S_o, we have

$$0 = S_o + \frac{1}{l}[(1 - \beta)k - 1]n + \sum_i \lambda_i C_i \tag{5.42}$$

and

$$0 = \beta_i \frac{k}{l} n - \lambda_i C_i, \quad i = 1, 2, \ldots, 6. \tag{5.43}$$

Solving for C_i, inserting the result into Eq. (5.42), and making use of Eq. (5.31) yields

$$0 = S_0 + \frac{(k - 1)}{l} n. \tag{5.44}$$

Thus for the situation where a source is present $n = lS_0/(1 - k)$, which gives a positive result only when $k < 1$, that is, the system is subcritical. In the absence of a source, Eq. (5.44) is satisfied with a time-independent neutron population, n, only if $k = 1$, that is, the system is critical. Moreover, if $k = 1$, then any value of n satisfies the equation. These are the same conditions for steady state solutions as shown in Fig. 5.1, and given by Eqs. (5.24) and (5.28). Thus the presence of delayed neutrons has no effect on the requirements for achieving steady state neutron distributions.

Reactivity Formulation

The time-dependent behavior of a reactor is very sensitive to small deviations of the multiplication about one. This behavior is highlighted by writing the kinetics equations in terms of the reactivity, defined as

$$\rho = \frac{k-1}{k}. \tag{5.45}$$

Hence,

$$\rho \begin{cases} >0 & \textit{supercritical} \\ =0 & \textit{critical} \\ <0 & \textit{subcritical}. \end{cases}$$

Designating the prompt generation time as

$$\Lambda = 1/k \tag{5.46}$$

and using the definitions of ρ and Λ, we may simplify the kinetics equations—that is, Eqs. (5.40) and (5.41)—to

$$\frac{d}{dt}n(t) = S(t) + \frac{(\rho-\beta)}{\Lambda}n(t) + \sum_i \lambda_i C_i(t) \tag{5.47}$$

and

$$\frac{d}{dt}C_i(t) = \frac{\beta_i}{\Lambda}n(t) - \lambda_i C_i(t), \quad i = 1, 2, \ldots, 6. \tag{5.48}$$

In the many cases where the reactivity is not large, and thus, $k \approx 1$, we can approximate $\Lambda \approx 1$ without appreciably affecting solutions to the equations.

The steady state condition given by Eq. (5.44) may also be expressed in terms of ρ and Λ by setting the derivatives in Eqs. (5.47) and (5.48) equal to zero: The precursor solution is $C_i = (\beta_i/\lambda_i\Lambda)n$. Thus for a subcritical system with $S(t) = S_o$, Eq. (5.47) yields $n = \Lambda S_o/|\rho|$, whereas for a critical system, $\rho = 0$, any positive value of n is a solution of Eq. (5.47), provided $S(t) = 0$. Finally, we note that in most systems $\lambda_i\Lambda/\beta_i \ll 1$, and therefore under steady state conditions $C_i \gg n$. Thus the number of neutron-emitting fission products in a reactor is much larger than the number of neutrons.

5.5   Step Reactivity Changes

We next consider what happens when a step change in reactivity is applied to an initially critical reactor that has been operating at

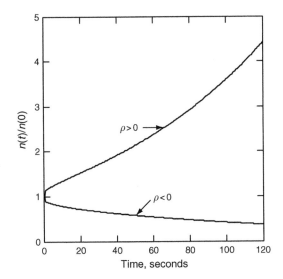

FIGURE 5.2 Neutron populations following reactivity insertions of $\pm 0.10\beta$.

steady state. The two curves in Fig. 5.2 show solutions to Eqs. (5.47) and (5.48) for positive and negative reactivity insertions. At first glance the curves appear similar to those in Fig. 5.1a. However, on further inspection we observe that the curves in Fig. 5.2 are more nuanced; this results from both prompt and delayed neutrons being taken into account. First—in less than a second—a prompt jump occurs in the neutron population. The jump is abrupt because it is controlled by the prompt neutron lifetime, here taken as $\Lambda = 50 \cdot 10^{-6}$ s, which is typical for a water-cooled reactor. There then follows the exponential growth or decay that appears similar for Fig. 5.1a. The exponential behavior displayed in Fig. 5.2 occurs quite slowly because the half-lives from the neutron-emitting fission products are the primary determinant the neutron population's growth or decay following the prompt jump.

The asymptotic behavior may be represented as $n(t) \propto \exp(t/T)$, where T is defined as the reactor period. The period is positive or negative depending on whether the reactor is super- or subcritical; it is the length of time required for the reactor power to increase or decrease by a factor of e and is arguably the most important quantity derivable from the kinetics equations.

Reactor Period

We determine the reactor period by returning to the kinetics equations and setting the source equal to zero. We then look for a solution of the seven equations set in the form $n(t) = A \exp(\omega t)$ and

$C_i(t) = B_i \exp(\omega t)$, where A, B_i, and ω are constants. Inserting these expressions into Eqs. (5.47) and (5.48) yields

$$\omega A = \frac{(\rho - \beta)}{\Lambda} A + \sum_i \lambda_i B_i \qquad (5.49)$$

and

$$\omega B_i = \frac{\beta_i}{\Lambda} A - \lambda_i B_i. \qquad (5.50)$$

Solving the second equation for B_i in terms of A and inserting the result into the first, we obtain—after canceling out the A's—the following:

$$\omega = \frac{(\rho - \beta)}{\Lambda} + \frac{1}{\Lambda} \sum_i \frac{\beta_i \lambda_i}{\omega + \lambda_i}. \qquad (5.51)$$

Next, we consolidate the terms on the right by making use of Eq. (5.31):

$$\omega = \frac{\rho}{\Lambda} - \frac{1}{\Lambda} \sum_i \frac{\beta_i}{\omega + \lambda_i} \omega. \qquad (5.52)$$

Finally, solving for ρ yields

$$\rho = \left(\Lambda + \sum_i \frac{\beta_i}{\omega + \lambda_i} \right) \omega. \qquad (5.53)$$

This is known as the inhour equation, since the units of ω are commonly taken as inverse hours.

The solution of Eq. (5.53) may be examined by graphing the right-hand side versus ω as Fig. 5.3 illustrates. By drawing a horizontal line for a specific value of ρ, we observe that there are seven solutions, say, $\omega_1 > \omega_2 \cdots > \omega_7$, regardless of whether the reactivity is positive or negative. Accordingly, the neutron population takes the form

$$n(t) = \sum_{i=1}^{7} A_i \exp(\omega_i t). \qquad (5.54)$$

Figure 5.3 indicates that for positive reactivity only ω_1 is positive. The remaining terms rapidly die away, yielding an asymptotic solution in the form

$$n(t) \simeq A_1 \exp(t/T), \qquad (5.55)$$

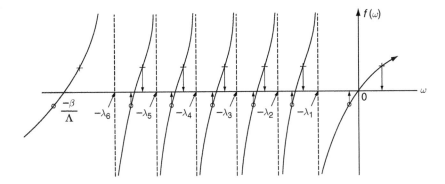

FIGURE 5.3 Solution of the in-hour equation (adapted from A. F. Henry, *Nuclear-Reactor Analysis*, 1975, by permission of the MIT Press).

where $T = 1/\omega_1$ is the reactor period. Figure 5.3 also shows that a negative reactivity leads to a negative period: All of the ω_i are negative, but $T = 1/\omega_1$ will die away more slowly than the others. Thus Eq. (5.55) is valid for negative as well as positive reactivities.

The plots in Fig. 5.2 are for reactivities of $\rho = \pm 0.1\beta$, using parameters for uranium-235 and $\Lambda = 50 \cdot 10^{-6}$ s. The prompt jump magnitude at the beginning of the curves is approximately $|A_1 - n(0)|$.

To determine the reactivity required to produce a given period—or vice versa—a plot of ρ vs T must be constructed using the delayed neutron data for a particular fissionable isotope or isotopes, and for a given prompt neutron lifetime. Figure 5.4 employs uranium-235 data in such a plot. Note that the very rapid decrease in the period takes place as ρ exceeds β. The condition $\rho = \beta$ defines prompt critical, for at that point the chain reaction is sustainable without delayed neutrons as indicated by the change in sign from negative to positive of the second term in Eq. (5.47).

As prompt criticality is exceeded, the distinction between the prompt jump and the reactor period vanishes, for now the prompt neutron lifetime rather than the delayed neutron half-lives largely determines the rate of exponential increase. Indeed, as prompt critical is approached, the period becomes so short that controlling the reactor by mechanical means such as the movement of control rods becomes exceeding difficult if not impossible. So important is it to avoid approaching prompt critical that reactivity is often measured in dollars, $\$ = \rho/\beta$, or in cents.

For negative reactivities there is an asymptotic limit to how fast the neutron population can be decreased. Note from Fig. 5.3 that the smallest negative period possible—that is, the fastest a reactor's power can be decreased after the initial prompt drop—is determined by the

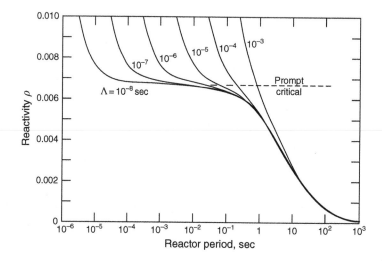

FIGURE 5.4   Reactivity vs reactor period for various prompt neutron generation times.

longest half-life of delayed neutrons or $T = 1/\lambda_1 = 0.693 t_{1/2}|_{max}$, which is approximately 56 s for a uranium-fueled reactor.

Figure 5.3 indicates that when the reactivity is small, so is the largest value of omega, ω_1. Suppose for a small reactivity we take $\omega_1 \ll \lambda_i$ for all λ_i. We may then eliminate omega from the denominator of Eq. (5.53), yielding

$$\rho = \left(\Lambda + \sum_i \frac{\beta_i}{\lambda_i} \right) \omega_1. \tag{5.56}$$

Using Eq. (5.34) to eliminate the summation, and solving for omega, we obtain $\omega_1 = \rho/(\Lambda + \beta/\lambda)$. Generally, $\Lambda \ll \beta/\lambda$, yielding

$$T \approx \beta/(\rho\lambda). \tag{5.57}$$

Thus for small reactivities—positive or negative—the reactor period is governed almost completely by the delayed neutron properties β and λ. This may seem surprising since delayed neutrons are such a small fraction of the fission neutrons produced. But when this small fraction is multiplied by the average half-life, $\beta/\lambda = 0.693\beta t_{1/2}$, a time substantially longer than the prompt neutron lifetime results.

Above prompt critical the period is very small, and thus ω_1 is very large. In this situation, we may take $\omega_1 \gg \lambda_i$ for all λ_i in Eq. (5.53), reducing it to $\rho = \Lambda\omega_1 + \beta$, or equivalently $\omega_1 = (\rho - \beta)/\Lambda$. Thus the reactor period is very short for $\rho > \beta$, for it is proportional to the prompt neutron generation time and independent of the delayed neutron half-lives:

$$T \approx \Lambda/(\rho - \beta). \tag{5.58}$$

Prompt Jump Approximation

The prompt jump that occurs following small step changes in reactivity may be utilized in making experimental determinations of reactor parameters using the rod drop and source jerk techniques. This jump is observed in Fig. 5.5, where we have magnified the initial part of the flux transients appearing in Fig. 5.2. Note that although the neutron population jumps rapidly at first, it then undergoes change much more slowly. The precursor concentration behaves much more sluggishly than the neutron population, changing hardly at all over the time it takes for the neutrons to undergo the initial jump up or down. This behavior stems from the long precursor half-lives compared to the prompt generation time; the precursor behavior governed by Eq. (5.48) is sluggish even when responding to a rapid change in neutron population since the decay constant multiplying

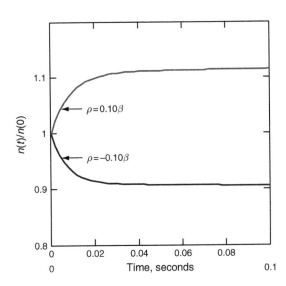

FIGURE 5.5   Prompt jump in neutron populations following reactivity insertions of $\pm 0.10\beta$.

C_i is small. In contrast, the neutron population at first responds very rapidly as a result of the small value of prompt neutron generation time Λ in the denominator of Eq. (5.47). After a short time has elapsed, however, the neutron response is slowed as a result of the more slowly changing value of the last term in Eq. (5.48). We can make use of this behavior to estimate the magnitudes of the prompt jump in the two following situations.

Rod Drop

Suppose we have a source-free critical reactor operating at steady state. According to Eq. (5.48) the precursor concentrations are related to the neutron population by

$$C_{io} = (\beta_i/\lambda_i\Lambda)n_o. \tag{5.59}$$

Immediately following the rod insertion the precursor concentration C_{io} will remain essentially unchanged for as long as $\lambda_i t \ll 1$. Thus the source-free form of Eq. (5.47) becomes

$$\frac{d}{dt}n(t) = -\frac{(|\rho| + \beta)}{\Lambda}n(t) + \frac{\beta}{\Lambda}n_o, \quad t \ll 1/\lambda_i, \tag{5.60}$$

where we have employed Eq. (5.31) and taken $\rho = -|\rho|$ to indicate a negative reactivity. We use the integrating factor technique detailed in Appendix A to obtain a solution. With an integrating factor of $\exp[(|\rho| + \beta)/\Lambda]$ and an initial condition of $n(0) = n_o$, we obtain after some simplification

$$n(t) = \frac{\beta}{(|\rho| + \beta)}n_o + \frac{|\rho|}{(|\rho| + \beta)}n_o e^{-\frac{1}{\Lambda}(|\rho|+\beta)t}. \tag{5.61}$$

Following decay of the second term, which is very rapid, we have for times substantially greater than the prompt generation time, but less than the half-lives of the delayed neutrons (i.e., $\Lambda/(|\rho| + \beta) \ll t \ll 1/\lambda_i$), a neutron population of approximately

$$n_1 = \frac{\beta}{(|\rho| + \beta)}n_o. \tag{5.62}$$

Thus the reactivity drop in dollars can be measured by taking the ratio of the neutrons immediately before to immediately after the control rod insertion:

$$|\rho|/\beta = \frac{n_o}{n_1} - 1. \tag{5.63}$$

Source Jerk

The source jerk is performed on a subcritical reactor, $\rho = -|\rho|$, containing a neutron source S_o. Because the system is in a steady state, Eq. (5.43) holds, and the initial condition for the experiment is $C_{io} = (\beta_i/\lambda_i\Lambda)n_o$. Following sudden removal (i.e., jerking) of the source, the neutron population undergoes a sharp negative jump. We can again model the transient using Eq. (5.47) with the source term set equal to zero, and $C_i(t)$ replaced with C_{io} for times $t \ll 1/\lambda_i$. Equation (5.60) once again results, with a solution given by Eq. (5.61), and Eqs. (5.62) and (5.63) remain valid in describing the relationship between reactivity, delayed neutron fraction, and the decrease in neutron population. Thus we see that either inserting a negative reactivity $\rho = -|\rho|$ into a critical reactor (the rod drop experiment) or removing the source from a subcritical system with reactivity $\rho = -|\rho|$ (the source jerk experiment) results in a measurable drop in the neutron population from n_o to n_1.

Rod Oscillator

A third experimental procedure may be analyzed in terms of Eqs. (5.47) and (5.48). However, in the rod oscillator experiment the equations may not be solved so simply. Laplace transforms or related techniques must be used to solve the equations in the presence of a sinusoidal reactivity of the form $\rho(t) = \rho_o \sin(\omega t)$. Here we simply state the result: If n_o is the initial neutron population, then after transients have died out the neutron population will oscillate as

$$n(t) = n_o[1 + A(\omega)\sin(\omega t) + \omega\rho_o/\Lambda], \qquad (5.64)$$

where $A(\omega)$ is a function of the frequency. If $\sin(\omega t)$ is averaged over time the sinusoidal term vanishes leaving $\bar{n} = n_o(1 + \omega\rho_o/\Lambda)$. Thus plotting $\bar{n}/n_o$ versus frequency determines ρ_o/Λ, the ratio of reactivity to prompt neutron generation time.

5.6 Prologue to Reactor Dynamics

Thus far our treatment of the time-dependent behavior of chain reactions has not accounted for thermal feedback effects. For this reason such treatments are frequently referred to as zero-power kinetics. If the energy created by fission, however, is large enough to cause the temperature of the system to rise, the material densities will then change. Because macroscopic cross sections are proportional to densities they too will change. Aside from the material density changes, additional feedback effects will result from the resonance cross sections, which broaden and flatten with increasing temperature as a result of the Doppler broadening discussed in Chapter 4. In thermal reactors the spectrum will also harden as a result of the temperature dependence of the Maxwell-Boltzmann distribution. Taken together these feedback effects will alter the parameters in the kinetics equations. By far the largest impact is on the reactivity. Consequently, reactor design must assure that under all operating conditions the feedback is negative for increases in temperature.

Negative feedback impacts the curves in Fig 5.2, for example, in the following way. When the neutron population becomes large enough for a temperature rise to occur, the curve will flatten if negative feedback is present, and then stabilize and possibly decrease with time. In Fig. 5.6 we have redrawn the positive reactivity curve of Fig 5.2 on a logarithmic scale, along with a curve for the same reactivity insertion but for which the effects of negative temperature feedback are included. Note that both curves initially follow the same period as indicated by the straight-line behavior on the logarithmic plot. But as the power becomes larger the curve with feedback becomes concave downward and stabilizes at a constant power. At this point the negative feedback has completely compensated for the initial reactivity insertion. In Chapter 9 we shall take up reactivity feedback in detail and examine its interaction with reactor kinetics to determine the transient behavior of power reactors.

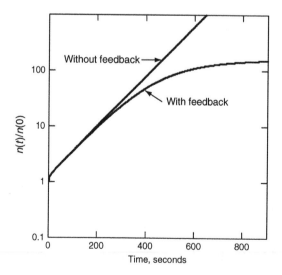

FIGURE 5.6   Effect of negative temperature feedback on neutron population following a reactivity insertion of $+0.10\beta$.

Bibliography

Bell, George I., and Samuel Glasstone, *Nuclear Reactor Theory*, Van Nostrand Reinhold, NY, 1970.

Duderstadt, James J., and Louis J. Hamilton, *Nuclear Reactor Analysis*, Wiley, NY, 1976.

Henry, Allen F., *Nuclear-Reactor Analysis*, MIT Press, Cambridge, MA, 1975.

Hetrick, David, *Dynamics of Nuclear Systems*, American Nuclear Society, 1993.

Keepin, G. R., *Physics of Nuclear Kinetics*, Addison-Wesley, Reading, MA, 1965.

Knief, Ronald A., *Nuclear Energy Technology: Theory and Practice of Commerical Nuclear Power*, McGraw-Hill, 1981.

Lewis, E. E., *Nuclear Power Reactor Safety*, Wiley, NY, 1977.

Ott, Karl O., and Robert J. Neuhold, *Introductory Nuclear Reactor Dynamics*, American Nuclear Society, 1985.

Schultz, M. A., *Control of Nuclear Reactors and Power Plants*, McGraw-Hill, NY, 1961.

Stacey, Weston M., *Nuclear Reactor Physics*, Wiley, NY, 2001.

Problems

5.1. a. What is l_∞ of 0.0253 eV thermal neutrons in graphite ($\Sigma_a = 0.273 \cdot 10^{-3}\,\text{cm}^{-1}$)?

    b. What is l_∞ of 1 MeV fast neutrons in iron ($\Sigma_a = 0.738 \cdot 10^{-3}\,\text{cm}^{-1}$)?

    (Note that a 0.0253 eV neutron has a speed of 2200 m/s.)

5.2. A power reactor is fueled with slightly enriched uranium. At the end of core life 30% of the power comes from the fissioning of the built up plutonium-239. Calculate the effective value of β at the beginning and at the end of core life; determine the percent increase or decrease.

5.3. At $t = 0$ there are no neutrons in a reactor. A neutron source is inserted into the reactor at $t = 0$ but then withdrawn at $t = 1$ min. Sketch the neutron population for $0 \le t \le 2$ min:

    a. For a subcritical reactor.
    b. For a critical reactor.

5.4. Suppose that a fissile material is discovered for which all of the neutrons are prompt. The neutron population is then governed by Eq. (5.22). Furthermore suppose that a reactor fueled with this material has a prompt neutron lifetime of 0.002 s.

a. If the reactor is initially critical, and there is no source present, what period should the reactor be put on if it is to triple its power in 10 s?

b. What is the reactivity ρ needed in part a?

5.5. Show that Eqs. (5.47) and (5.48) result from inserting the definitions of ρ and Λ into Eqs. (5.40) and (5.41).

5.6. A thermal reactor fueled with uranium operates at 1.0 W. The operator is to increase the power to 1.0 kW over a 2 hour span of time.

a. What reactor period should she put the reactor on?

b. How many cents of reactivity must be present to achieve the period in part a?

5.7. A thermal reactor fueled with uranium operates at 1.0 W. The operators put in on a 15-minute period. How long will the reactor take to reach a power of 1.0 MW?

5.8. Show that Eq. (5.53) follows from Eq. (5.51).

5.9. Find the periods for reactors fueled by uranium-235, plutonium-239, and uranium-233 if

a. One cent of reactivity is added to the critical systems.

b. One cent of reactivity is withdrawn from the critical systems.

5.10. The one delayed group approximation results from lumping all six precursors into one, $C(t) = \sum_{i=1}^{6} C_i(t)$, and replacing the λ_i by the average value defined by Eq. (5.34). Equations (5.47) and (5.48) then reduce to the one delayed group equations:

$$\frac{d}{dt} n(t) = S(t) + \frac{(\rho - \beta)}{\Lambda} n(t) + \lambda C(t)$$

and

$$\frac{d}{dt} C(t) = \frac{\beta}{\Lambda} n(t) - \lambda C(t).$$

With these equations, consider a critical reactor that is initially operating with a neutron population of $n(0)$ and for which $S(t) = 0$. At $t = 0$ a step reactivity change ρ is made. Using the assumptions that

$$\frac{1}{\lambda \Lambda} |\beta - \rho| \gg 1 \quad \text{and} \quad \frac{\beta}{\lambda \Lambda} \gg 1$$

a. Show that

$$n(t) = n(0) \left\{ \frac{\rho}{\rho - \beta} \exp\left[\frac{\rho - \beta}{\Lambda} t\right] + \frac{\beta}{\beta - \rho} \exp\left[\frac{\lambda\rho}{\beta - \rho} t\right] \right\}.$$

b. Show that for long times the solution is independent of λ when $\rho > \beta$ and independent of Λ when $0 < \rho < \beta$.

c. Taking $\beta = 0.007$, $\Lambda = 5 \cdot 10^{-5}$ s, and $\lambda = 0.08 \, \text{s}^{-1}$, make a graph of the reactivity in dollars vs the reactor period for reactivities between -2 dollars and $+2$ dollars. Indicate the region or regions on the graph where you expect the results from part a to be poor.

5.11. Using the kinetics equations with one delayed group from problem 5.10,

a. Find the reactor period when $\rho = \beta$, simplifying your answer by assuming $\lambda\Lambda/\beta \ll 1$.

b. Calculate the reactor period for $\beta = 0.007$, $\Lambda = 5 \cdot 10^{-5}$ s, and $\lambda = 0.08 \, \text{s}^{-1}$.

5.12. By differentiating the kinetics equations with one group of delayed neutrons (given in problem 5.10) and then letting $\Lambda \to 0$,

a. Show that $C(t)$ can be eliminated to obtain

$$\frac{d}{dt} n(t) = \frac{\lambda}{\beta - \rho} \left(\rho + \frac{1}{\lambda} \frac{d\rho}{dt} \right) n(t),$$

which is referred to as the zero lifetime or prompt jump approximation.

b. For a step change in reactivity of $|\rho| \ll \beta$, find the zero lifetime approximation to the reactor period.

5.13. A critical reactor operates at a power level of 80 W. Dropping a control rod into the core causes the flux to undergo a sudden decrease to 60 W. How many dollars is the control rod worth?

5.14. Estimate the period of the reactor from the curve without feedback in Fig. 5.6. Suppose you wanted to put the reactor on a period of one minute. What reactivity would you insert?

5.15. Solve the kinetics equations numerically with one group of delayed neutrons, given in problem 5.10, using the data $\beta = 0.007$, $\Lambda = 5 \cdot 10^{-5}$ s, and $\lambda = 0.08 \, \text{s}^{-1}$,

a. For a step insertion of $+0.25$ dollars between 0 and 5 s.

b. For a step insertion of -0.25 dollars between 0 and 5 s.

Assume that the reactor is initially in the critical state, and plot $n(t)/n(0)$ for each case.

5.16. Repeat problem 5.15 with six groups of delayed neutrons, employing the data for uranium-235 given in Table 5.1 and a prompt generation time of 50×10^{-6} s.

5.17. Consider a subcritical reactor described by the pair of kinetics equations in problem 5.10, and with $\beta = 0.007$, $\Lambda = 5 \cdot 10^{-5}$ s, and $\lambda = 0.08 \, \mathrm{s}^{-1}$. The subcritical system is in steady state equilibrium with reactivity $\rho = -10$ cents and a time-independent source S_o. Then, at $t = 0$ the source is removed. Determine $n(t)$ for $t \geq 0$ and plot your result, normalized to S_o.

5.18. Solve Eqs. (5.47) and (5.48) using the uranium-235 data from Table 5.1 and a prompt generation time of 50×10^{-6} s for the following three ramp insertions.

   a. $\rho(t) = 0.25\beta t$.
   b. $\rho(t) = 0.5\beta t$.
   c. $\rho(t) = 1.0\beta t$.

   Normalizing your results to $n(0) = 1$, make linear plots and for each case determine: (1) at what time does the neutron population reach $n(t)/n(0) = 1000$ and (2) what is the value of $n(t)/n(0)$ at the point in time when the system reaches prompt critical?

5.19. Suppose that a critical reactor is operating at a steady state level with a neutron population of n_o. You are to add reactivity such that the neutron population will increase linearly with time: $n(t) = n(0)(1 + \vartheta t)$ where ϑ is a constant. Using the one delayed group kinetics equations from problem 5.10,

   a. Determine the time-dependent reactivity $\rho(t)$ that you should add to the reactor to achieve the linear increase.
   b. Sketch $\rho(t)$ from part a.

CHAPTER 6

Spatial Diffusion of Neutrons

6.1 Introduction

Chapter 5 dealt with time-dependent behavior of nuclear reactors, and in earlier chapters we examined the importance of the energy spectrum of neutrons in determining the multiplication and other reactor properties. Thus far, we have not dealt with the spatial distributions of neutrons, other than the treatment in Chapter 4 of their distributions between fuel, coolant, and/or moderator within lattice cells. We have characterized the effects of the global distribution of neutrons simply by a nonleakage probability, which as stated earlier increases toward a value of one as the reactor core becomes larger. In this chapter and the next we examine the spatial migration of neutrons, not only to obtain an explicit expression for the nonleakage probability, but also to understand the relationships between reactor size, shape, and criticality, and to determine the spatial flux distributions within power reactors.

In treating the spatial distributions of neutrons we will consider a monoenergetic or one energy group model. That means that the neutrons' flux and cross sections have already been averaged over energy. Likewise, when dealing with reactor lattices we assume that the flux and cross sections have been spatially averaged over the lattice cell. Thus we are dealing only with the global variations of the neutron distribution, and not the spatial fluctuations with the periodicity of the lattice cell pitch.

The neutron diffusion equation provides the most straightforward approach to determining spatial distributions of neutrons. In this chapter we first derive the diffusion equation and its associated boundary conditions. We then apply diffusion theory to problems in nonmultiplying media, limiting our attention in this chapter to problems in highly idealized one-dimensional geometries—first plane and then spherical geometry—for they offer the clearest introduction to the solution techniques. We then proceed to examine the behavior in spherical systems that contain fissionable material but which are subcritical.

We conclude the chapter with an approach to criticality by increasing either the value of k_∞ or the size of the reactor, obtaining a criticality equation at the chapter's end. We thus set the stage for Chapter 7 where we take up the important issues related to the spatial distributions of flux and power in the finite cylindrical cores of power reactors.

6.2   The Neutron Diffusion Equation

To derive the neutron diffusion equation, we first set forth the neutron balance condition for an incremental volume. We then employ Fick's law to arrive at the diffusion equation.

Spatial Neutron Balance

To begin, consider the neutron balance within an infinitesimal volume element $dV = dxdydz$ centered at the point $\vec{r} = (x, y, z)$ as shown in Fig. 6.1. Under steady state conditions neutron conservation requires that

$$\begin{pmatrix} \text{neutrons leaking} \\ \text{out of } dxdydz/\text{s} \end{pmatrix} + \begin{pmatrix} \text{neutrons absorbed} \\ \text{in } dxdydz/\text{s} \end{pmatrix}$$

$$= \begin{pmatrix} \text{source neutrons} \\ \text{emitted in } dxdydz/\text{s} \end{pmatrix} + \begin{pmatrix} \text{fission neutrons} \\ \text{produced in } dxdydz/\text{s} \end{pmatrix}. \tag{6.1}$$

The leakage is the sum of the net number of neutrons passing out through the six surfaces of the cubical volume. To express the leakage

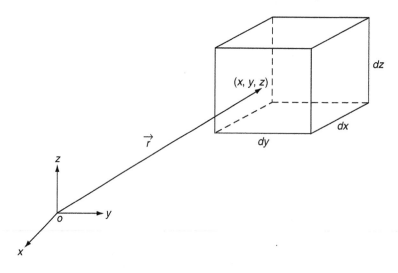

FIGURE 6.1   Control volume for neutron balance equation.

quantitatively, we define the three components of the neutron current. Let $J_x(x, y, z)$ be the net number of neutrons/cm^2/s passing though the y–z plane in the positive x direction at (x, y, z). Similarly, let $J_y(x, y, z)$ and $J_z(x, y, z)$ be the net numbers of neutrons/cm^2/s passing though the x–z and x–y planes, respectively. For the cubical volume, the net number of neutrons passing out of the cube through the front, right, and top faces is then $J_x(x + \frac{1}{2}dx, y, z)dydz$, $J_y(x, y + \frac{1}{2}dy, z)dxdz$, and $J_z(x, y, z + \frac{1}{2}dz)dxdy$. Likewise, the net number of neutrons passing out of the back, left, and bottom faces is $-J_x(x - \frac{1}{2}dx, y, z)dydz$, $-J_y(x, y - \frac{1}{2}dy, z)dxdz$, and $-J_z(x, y, z - \frac{1}{2}dz)dxdy$. Thus the net leakage per second from the cube is

$$\begin{pmatrix} \text{neutrons leaking} \\ \text{out of } dxdydz/\text{s} \end{pmatrix} = [J_x(x + \frac{1}{2}dx, y, z) - J_x(x - \frac{1}{2}dx, y, z)]dydz$$
$$+ [J_y(x, y + \frac{1}{2}dy, z) - J_y(x, y - \frac{1}{2}dy, z)]dxdz$$
$$+ [J_z(x, y, z + \frac{1}{2}dz) - J_z(x, y, z - \frac{1}{2}dz)]dxdy. \tag{6.2}$$

Since by definition the partial derivative is

$$\lim_{dx \to 0}[J_x(x + \frac{1}{2}dx, y, z) - J_x(x - \frac{1}{2}dx, y, z)]/dx \equiv \frac{\partial}{\partial x}J_x(x, y, z), \quad (6.3)$$

and similarly in the y and z directions, we multiply and divide the three terms on the right of Eq. (6.2) by dx, dy, and dz, respectively, and then take the limits to obtain

$$\begin{pmatrix} \text{neutrons leaking} \\ \text{out of } dxdydz/\text{s} \end{pmatrix} = \left[\frac{\partial}{\partial x}J_x(x, y, z) + \frac{\partial}{\partial y}J_y(x, y, z) \right.$$
$$\left. + \frac{\partial}{\partial z}J_z(x, y, z) \right] dxdydz. \tag{6.4}$$

The remaining terms are

$$\begin{pmatrix} \text{neutrons absorbed} \\ \text{in } dxdydz/\text{s} \end{pmatrix} = \Sigma_a(x, y, z)\phi(x, y, z)dxdydz, \tag{6.5}$$

$$\begin{pmatrix} \text{source neutrons} \\ \text{emitted in } dxdydz/\text{s} \end{pmatrix} = s'''(x, y, z)dxdydz, \tag{6.6}$$

and

$$\begin{pmatrix} \text{fission neutrons} \\ \text{produced in } dxdydz/\text{s} \end{pmatrix} = \nu\Sigma_f(x, y, z)\phi(x, y, z)dxdydz. \tag{6.7}$$

Inserting Eqs. (6.4) through (6.7) into Eq. (6.1) and eliminating $dxdydz$ then yields

$$\frac{\partial}{\partial x} J_x(\vec{r}) + \frac{\partial}{\partial y} J_y(\vec{r}) + \frac{\partial}{\partial z} J_z(\vec{r}) + \Sigma_a(\vec{r})\phi(\vec{r}) = s'''(\vec{r}) + \nu\Sigma_f(\vec{r})\phi(\vec{r}). \quad (6.8)$$

For compactness, we may write the current in vector form,

$$\vec{J}(\vec{r}) = \hat{i}J_x(\vec{r}) + \hat{j}J_y(\vec{r}) + \hat{k}J_z(\vec{r}), \quad (6.9)$$

and use the definition of the gradient, $\vec{\nabla} = \hat{i}\frac{\partial}{\partial x} + \hat{j}\frac{\partial}{\partial y} + \hat{k}\frac{\partial}{\partial z}$, to write Eq. (6.8) as

$$\vec{\nabla} \cdot \vec{J}(\vec{r}) + \Sigma_a(\vec{r})\phi(\vec{r}) = s'''(\vec{r}) + \nu\Sigma_f(\vec{r})\phi(\vec{r}). \quad (6.10)$$

Diffusion Approximation

Equation (6.10) is a statement of neutron balance. In most of the circumstances that we will consider, we may reasonably apply Fick's law—or more precisely, Fick's approximation—to relate the current to the flux:

$$\vec{J}(\vec{r}) = -D(\vec{r})\vec{\nabla}\phi(\vec{r}), \quad (6.11)$$

where D is referred to as the diffusion coefficient. Inserting Eq. (6.11) into Eq. (6.10) then yields the neutron diffusion equation:

$$-\vec{\nabla} \cdot D(\vec{r})\vec{\nabla}\phi(\vec{r}) + \Sigma_a(\vec{r})\phi(\vec{r}) = s'''(\vec{r}) + \nu\Sigma_f(\vec{r})\phi(\vec{r}). \quad (6.12)$$

More advanced neutron transport techniques are employed in Appendix C to evaluate the diffusion coefficient as

$$D = 1/3\Sigma_{tr}, \quad (6.13)$$

where the transport cross section is defined as $\Sigma_{tr} = \Sigma_t - \bar{\mu}\Sigma_s$, where $\bar{\mu}$ is the average scattering angle. For isotropic scattering in the laboratory system, $\bar{\mu} = 0$ and thus the transport cross section reduces to the total cross section, Σ_t. Averaging the diffusion coefficient for mixtures of materials is accomplished by first using the techniques of Chapter 2 or 4 to average Σ_{tr} and then applying Eq. (6.13).

   In what follows, we first solve the diffusion equation for non-multiplying media in simple one-dimensional geometries, and stipulate the boundary conditions that apply to it under a variety of circumstances. Before treating multiplying media, we return to examine the circumstances under which the diffusion approximation is valid and the meaning of the diffusion length, which following its introduction makes ubiquitous appearances in the solutions of diffusion problems.

6.3 Nonmultiplying Systems—Plane Geometry

We consider first the case of a uniform medium with no fissionable material, that is, a nonmultiplying medium. Thus $\Sigma_f = 0$, and D and Σ_a are constants, allowing Eq. (6.12) to be reduced to

$$-\nabla^2 \phi(\vec{r}) + \frac{1}{L^2}\phi(\vec{r}) = \frac{1}{D}s'''(\vec{r}), \tag{6.14}$$

where $\nabla^2 \equiv \dfrac{\partial^2}{\partial x^2} + \dfrac{\partial^2}{\partial y^2} + \dfrac{\partial^2}{\partial z^2}$, and the diffusion length is defined by

$$L = \sqrt{D/\Sigma_a}, \tag{6.15}$$

which has units of length.

Source Free Example

To illustrate the significance of the diffusion length we consider a simple problem in plane geometry, that is, where the flux varies so slowly in y and z that it can be ignored, allowing us to eliminate the y and z derivatives from ∇^2. We also set the source to zero, simplifying Eq. (6.14) to

$$\frac{d^2}{dx^2}\phi(x) - \frac{1}{L^2}\phi(x) = 0. \tag{6.16}$$

We look for a solution of the form

$$\phi(x) = C\exp(\kappa x) \tag{6.17}$$

by substituting this expression into Eq. (6.16). The result is

$$C(\kappa^2 - 1/L^2)\exp(\kappa x) = 0. \tag{6.18}$$

The left-hand side is equal to zero for all x if $C = 0$, but that is unacceptable because it would cause the entire solution to vanish. Instead, we take

$$\kappa = \pm 1/L. \tag{6.19}$$

Thus there are two possible solutions, and the flux takes the form

$$\phi(x) = C_1 \exp(x/L) + C_2 \exp(-x/L). \tag{6.20}$$

To determine the coefficients C_1 and C_2 we must apply boundary conditions. Suppose that we want to solve for the problem domain that is a semi-infinite medium occupying the space $0 \le x \le \infty$. Further assume that neutrons are supplied from the left with sufficient intensity to provide a known flux ϕ_o at $x=0$, that is, $\phi(0) = \phi_o$. If no neutrons are entering from the right, then all of the neutrons entering from the left will eventually be absorbed as they diffuse to the right, requiring that $\phi(\infty) = 0$. We thus have the two needed boundary conditions. Inserting them into Eq. (6.20), we have

$$\phi_o = C_1 \exp(0/L) + C_2 \exp(-0/L), \tag{6.21}$$

and

$$0 = C_1 \exp(\infty/L) + C_2 \exp(-\infty/L). \tag{6.22}$$

Because $\exp(\infty) = \infty$ and $\exp(-\infty) = 0$ the second equation is satisfied only if $C_1 = 0$. Then because $\exp(0) = 1$ we obtain $C_2 = \phi_o$, and the solution is

$$\phi(x) = \phi_o \exp(-x/L). \tag{6.23}$$

Uniform Source Example

We next examine the case where there is a uniform source $s'''(\vec{r}) \to s_o'''$. In plane geometry, Eq. (6.14) reduces to

$$-\frac{d^2}{dx^2}\phi(x) + \frac{1}{L^2}\phi(x) = \frac{1}{D}s_o'''. \tag{6.24}$$

Solutions to problems containing source terms are divided into general and particular solutions:

$$\phi(x) = \phi_g(x) + \phi_p(x), \tag{6.25}$$

where ϕ_g, the general solution (sometimes also called the homogeneous solution), satisfies Eq. (6.24) with the source set to zero. Thus it is identical in form to the solution of Eq. (6.16), given by Eq. (6.20). The particular solution, ϕ_p, must satisfy Eq. (6.24) with the source present. For a uniform source the particular solution is a constant, since its derivative is zero. Thus the particular solution for Eq. (6.25) is

$$\phi_p(x) = \frac{L^2}{D}s_o''' = \frac{1}{\Sigma_a}s_o'''. \tag{6.26}$$

Substituting this expression along with Eq. (6.20) into Eq. (6.25) yields a solution of Eq. (6.24):

$$\phi(x) = C_1\exp(x/L) + C_2\exp(-x/L) + \frac{1}{\Sigma_a}s_0'''. \tag{6.27}$$

Once again we need two boundary conditions to specify the constants C_1 and C_2. Suppose the uniform source is distributed throughout a slab extending between $-a \le x \le a$ and that we specify the two boundary conditions as the flux vanishing at the slab surfaces: $\phi(\pm a) = 0$. The conditions on C_1 and C_2 then follow from Eq. (6.27):

$$0 = C_1\exp(a/L) + C_2\exp(-a/L) + \frac{1}{\Sigma_a}s_0''', \tag{6.28}$$

and

$$0 = C_1\exp(-a/L) + C_2\exp(a/L) + \frac{1}{\Sigma_a}s_0'''. \tag{6.29}$$

These equations are easily solved to show that $C_1 = C_2 = -(e^{a/L} + e^{-a/l})^{-1}s_o'''/\Sigma_a$. Thus the solution is

$$\phi(x) = \left(1 - \frac{\cosh(x/L)}{\cosh(a/L)}\right)\frac{s_o'''}{\Sigma_a}, \tag{6.30}$$

where we have used the hyperbolic cosine defined in Appendix A.

The uniform source results in the simplest form of a particular solution—a constant. Problems involving space-dependent source distributions can present more of a challenge in finding the particular solution. One such situation is included as problem 6.12.

6.4 Boundary Conditions

The problems introduced above indicate some general procedures used to solve the diffusion equation, which is a second order differential equation. In one-dimensional problems solutions contain two arbitrary constants. Two boundary conditions are needed to determine these coefficients. The conditions that we have used thus far are on the flux itself. Often other conditions more accurately represent the physical situation at hand. In deriving such conditions, a particularly useful concept is that of partial currents. Recall that the current $J_x(x)$ is the net number of neutrons/cm^2/s crossing a plane perpendicular to the x axis in the positive x direction. We may divide it into the partial currents, $J_x^+(x)$ and $J_x^-(x)$, of the neutrons traveling in the positive and negative x directions, respectively. Thus we have

$$J_x(x) = J_x^+(x) - J_x^-(x). \tag{6.31}$$

As detailed in Appendix C, neutron transport theory may be employed to show that in the diffusion approximation

$$J_x^{\pm}(x) = \frac{1}{4}\phi(x) \pm \frac{1}{2}J_x(x), \tag{6.32}$$

or employing Eq. (6.11) to eliminate the current,

$$J_x^{\pm}(x) = \frac{1}{4}\phi(x) \mp \frac{1}{2}D\frac{d}{dx}\phi(x). \tag{6.33}$$

Vacuum Boundaries

Suppose we have a surface across which no neutrons are entering. This would be the case if a vacuum containing no neutron sources extended to infinity. We refer to this as a vacuum boundary. If the boundary is on the left, say, at x_l, the condition is $J_x^+(x_l) = 0$. If it is on the right, say, at x_r, the condition is $J_x^-(x_r) = 0$. Using the definition of the partial current, we may write the condition on the right as

$$0 = \frac{1}{4}\phi(x_r) - \frac{1}{2}D\left|\frac{d}{dx}\phi(x)\right|_{x_r}, \tag{6.34}$$

where since the derivative is negative on the right, for clarity we have taken minus its absolute value. Recalling that for isotropic scattering $D = 1/(3\Sigma_t)$ and that the mean free path is defined as $\lambda = 1/\Sigma_t$, we may rewrite the vacuum boundary condition as

$$\phi(x_r)\left/\left|\frac{d}{dx}\phi(x)\right|_{x_r}\right. = \frac{2}{3}\lambda. \tag{6.35}$$

Figure 6.2 provides a visual interpretation of the condition. If the flux is extrapolated linearly according to Eq. (6.35) it will go to zero at a distance of $2/3\lambda$ outside the boundary; thus we refer to $2/3\lambda$ as the extrapolation distance.

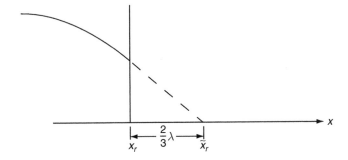

FIGURE 6.2   Neutron flux extrapolation distance at a vacuum boundary.

Frequently, when vacuum boundaries are encountered, simply adjusting the dimensions and using a zero flux boundary condition is more straightforward. Thus for a problem with left and right vacuum boundaries at x_l and x_r, we would take $\phi(x_l - 2/3\lambda) = 0$ and $\phi(x_r + 2/3\lambda) = 0$, respectively. For brevity, we shall frequently denote such an extrapolated boundary simply by adding a tilde to the spatial dimension. Thus $\phi(\tilde{x}_r) = 0$ and $\phi(\tilde{x}_l) = 0$. In situations where the problem dimensions are large when measured in λ, a common practice consists of ignoring the small correction for the extrapolation distance and simply taking $\phi(x_r) \approx 0$ and $\phi(x_l) \approx 0$.

Reflected Boundaries

If the net current J_x is known at a boundary, it also may be used as a condition. Current conditions appear most frequently as a result of problem symmetry. Suppose, as in the uniform source problem described above, the solution is symmetric about $x = 0$, such that the net number of neutrons crossing the plane at $x = 0$ vanishes. Then we may solve the problem only for $x > 0$ using the boundary condition on the left as

$$J_x(0) = -D\frac{d}{dx}\phi(x)\bigg|_{x=0} = 0, \qquad (6.36)$$

or simply

$$\frac{d}{dx}\phi(x)\bigg|_{x=0} = 0.$$

Surface Sources and Albedos

Partial currents are particularly useful in specifying boundary conditions in situations where surfaces are bombarded by neutrons. To illustrate, consider a source free solution, such as Eq. (6.23). For this solution to hold, neutrons must be being supplied from the left. We can relate the boundary flux ϕ_o to a surface source of s'' neutrons/cm^2/s by noting that the number of neutrons entering from the left is just

$$J_x^+(0) = s''. \tag{6.37}$$

Thus combining Eqs. (6.23) and (6.33) to obtain the partial current,

$$J_x^+(x) = \left(\frac{1}{4} + \frac{D}{2L}\right)\phi_o \exp(-x/L), \tag{6.38}$$

we can use Eq. (6.37) to eliminate ϕ_o and write Eq. (6.38) directly in terms of the surface source s'':

$$\phi(x) = \left(\frac{1}{4} + \frac{D}{2L}\right)^{-1} s'' \exp(-x/L). \tag{6.39}$$

Of the surface source neutrons entering the diffusing medium, some fraction will make scattering collisions and then come back out. The ratio of exiting to entering neutrons is termed the albedo. We express it in terms of the partial currents as

$$\alpha = \frac{J_x^-(0)}{J_x^+(0)}. \tag{6.40}$$

The albedo for a semi-infinite medium results from inserting Eq. (6.33) into this expression and then using Eq. (6.23) to evaluate the partial currents at $x=0$. The result is

$$\alpha = (1 - 2D/L)/(1 + 2D/L). \tag{6.41}$$

Thus for the semi-infinite medium, $\alpha s''$ neutrons/cm^2/s will be reemitted from the surface, while the remaining neutrons, that is, $(1 - \alpha)s''$ neutrons/cm^2/s, will be absorbed within the medium. Equation (6.40) is applicable at any surface with $J_x^-(0)$ interpreted as the outgoing and $J_x^+(0)$ as the incoming partial current.

Interface Conditions

If more than one region is present, say, with different cross sections, the diffusion equation solution will contain two arbitrary constants for each region. Thus two conditions are required at each interface. They are that both the flux and the net current must be continuous. Thus for an interface at x_o we have

$$\phi(x_{o-}) = \phi(x_{o+}) \tag{6.42}$$

and

$$D(x_{o-})\frac{d}{dx}\phi(x)|_{x_{o-}} = D(x_{o+})\frac{d}{dx}\phi(x)|_{x_{o+}}, \tag{6.43}$$

where x_{o+} and x_{o-} indicate evaluation immediately to the right and left of x_o. An exception to Eq. (6.43) occurs if there is a localized source present. If an infinitely thin surface source emits s''_{pl} neutrons/cm^2/s along the interface, the neutron balance at the interface is $J(x_{o-}) + s''_{pl} = J(x_{o+})$ or more explicitly

$$-D(x_{o-})\frac{d}{dx}\phi(x)|_{x_{o-}} + s''_{pl} = -D(x_{o+})\frac{d}{dx}\phi(x)|_{x_{o+}}. \tag{6.44}$$

Boundary Conditions in Other Geometries

The foregoing boundary conditions are applicable in cylindrical and spherical as well as Cartesian geometries: x is simply replaced by the direction normal to the surface. In spherical or cylindrical geometries, however, boundary conditions at the origin or centerline must be treated somewhat differently. For a point source emitting s_p neutrons/s at the center of a sphere, $s_p = \lim_{r\to 0}[4\pi r^2 J_r(r)]$, and for a line source emitting s'_l neutrons/cm/s along the centerline of a cylinder, $s'_l = \lim_{r\to 0}[2\pi r J_r(r)]$. If no such sources are present then these two conditions are equivalent simply to requiring that the flux be finite at $r = 0$. We illustrate the condition at the origin for spherical geometry in the following section, but defer the treatment of cylindrical geometry, which requires the introduction of Bessel functions, to the following chapter.

6.5   Nonmultiplying Systems—Spherical Geometry

We consider two spherical geometry problems. In the first a point source of neutrons is located in an infinite medium. The second is a two region problem with a distributed source. Both demonstrate boundary conditions at the origin, and the second also extends techniques introduced thus far to neutron diffusion problems to include the treatment of interface as well as boundary conditions.

   We begin by replacing ∇^2 in Eq. (6.14) by its one-dimensional spherical form found in Appendix A:

$$-\frac{1}{r^2}\frac{d}{dr}r^2\frac{d}{dr}\phi(r) + \frac{1}{L^2}\phi(r) = \frac{1}{D}s'''(r). \tag{6.45}$$

Point Source Example

Assume that a source of neutrons of strength s_p is concentrated at the origin of an infinite medium extending from $r = 0$ to $r = \infty$. We apply Eq. (6.45) for $r > 0$ with $s'''(r) = 0$:

$$\frac{1}{r^2}\frac{d}{dr}r^2\frac{d}{dr}\phi(r) - \frac{1}{L^2}\phi(r) = 0. \tag{6.46}$$

If we make the substitution

$$\phi(r) = \frac{1}{r}\psi(r), \tag{6.47}$$

however, the equation simplifies to

$$\frac{d^2}{dr^2}\psi(r) - \frac{1}{L^2}\psi(r) = 0, \tag{6.48}$$

which has the same form as Eq. (6.16) with x replaced by r. Thus we look for a solution of the form:

$$\psi(r) = C\exp(\kappa r), \tag{6.49}$$

and use the identical procedure as in Eqs. (6.16) through (6.20) to obtain

$$\psi(r) = C_1\exp(r/L) + C_2\exp(-r/L). \tag{6.50}$$

Thus from Eq. (6.47),

$$\phi(r) = \frac{C_1}{r}\exp(r/L) + \frac{C_2}{r}\exp(-r/L). \tag{6.51}$$

Next we employ the boundary conditions to determine C_1 and C_2. Infinitely far from the source at $r = \infty$ the flux must go to zero: $\phi(\infty) = 0$. Thus $C_1 = 0$, for otherwise we would have $\phi(\infty) \to \infty$. The boundary condition at the origin is a bit more subtle. In the limit as $r \to 0$ the current

$$J_r(r) = -D\frac{d}{dr}\phi(r) \tag{6.52}$$

emerging from a small sphere, with surface area $= 4\pi r^2$, must just be equal to the source strength. Hence

$$s_p = \lim_{r\to 0} 4\pi r^2 J_r(r). \tag{6.53}$$

Combining Eqs. (6.51) through (6.53), we find the flux distribution from a point source to be

$$\phi(r) = \frac{s_p}{4\pi Dr}\exp(-r/L). \tag{6.54}$$

Clearly all of the neutrons produced by the point source must be absorbed in the infinite medium. Taking an incremental volume as $dV = 4\pi r^2 dr$, we may show that

$$\int_{all\,space} \Sigma_a \phi(r) dV = s_p. \tag{6.55}$$

Two Region Example

The second problem we consider has two regions and a distributed source. It illustrates the treatment of the boundary condition at the origin as well as interface conditions. Suppose a sphere of radius R with material properties D and Σ_a contains a uniform source s_o'''. The sphere is surrounded by a second source free medium with properties $\hat{D}$ and $\hat{\Sigma}_a$ that extends to $r = \infty$. Our objective is to determine the neutron flux.

For this problem, we obtain the following two differential equations from Eq. (6.45):

$$-\frac{1}{r^2}\frac{d}{dr}r^2\frac{d}{dr}\phi(r) + \frac{1}{L^2}\phi(r) = \frac{1}{D}s_o''', \quad 0 \le r < R, \tag{6.56}$$

and

$$-\frac{1}{r^2}\frac{d}{dr}r^2\frac{d}{dr}\phi(r) + \frac{1}{\hat{L}^2}\phi(r) = 0, \quad R < r \le \infty. \tag{6.57}$$

Within the sphere we must superimpose general and particular solutions for Eq. (6.56):

$$\phi(r) = \phi_g(r) + \phi_p(r), \quad 0 \le r < R. \tag{6.58}$$

For a uniform source the particular solution is a constant. Thus

$$\phi_p = \frac{L^2}{D}s_o''' = \frac{s_o'''}{\Sigma_a}. \tag{6.59}$$

The general solution satisfies Eq. (6.46), and following the same transformation of variables we obtain Eq. (6.51) for $\phi_g(r)$. Thus inserting ϕ_g and ϕ_p into Eq. (6.58) we have

$$\phi(r) = \frac{C_1}{r}\exp(r/L) + \frac{C_2}{r}\exp(-r/L) + \frac{s_o'''}{\Sigma_a}, \quad 0 \le r < R. \tag{6.60}$$

Equations (6.46) and (6.57) are identical in form. Thus the solution to Eq. (6.57) has the same form as Eq. (6.51):

$$\phi(r) = \frac{C_1'}{r} \exp(r/\widehat{L}) + \frac{C_2'}{r} \exp(-r/\widehat{L}), \quad R < r \leq \infty, \qquad (6.61)$$

where $\widehat{L} = \sqrt{\widehat{D}/\widehat{\Sigma}_a}$.

The four arbitrary constants in Eqs. (6.60) and (6.61) are determined from the configuration's four boundary and interface conditions:

#1. $0 < \phi(0) < \infty$,   #3. $\phi(R_-) = \phi(R_+)$,

#2. $\phi(\infty) = 0$,   #4. $D\dfrac{d}{dr}\phi(r)|_{R_-} = \widehat{D}\dfrac{d}{dr}\phi(r)|_{R_+}$.   (6.62)

Applying boundary condition 1 by taking the limit of Eq. (6.60) as $r \to 0$, we see that the flux will remain finite only if $C_2 = -C_1$. Then using the definition of the hyperbolic sine defined in Appendix A, Eq. (6.60) reduces to

$$\phi(r) = \frac{2C_1}{r}\sinh(r/L) + \frac{s_0'''}{\Sigma_a}, \quad 0 \leq r < R. \qquad (6.63)$$

We next apply condition 2 to Eq. (6.61). Since the first term becomes infinite, but the second vanishes as $r \to \infty$, the condition is met if $C_1' = 0$:

$$\phi(r) = \frac{C_2'}{r}\exp(-r/\widehat{L}), \quad R < r \leq \infty. \qquad (6.64)$$

Finally we apply interface conditions 3 and 4 to obtain the remaining arbitrary coefficients:

$$\frac{2C_1}{R}\sinh(R/L) + \frac{s_0'''}{\Sigma_a} = \frac{C_2'}{R}\exp(-R/\widehat{L}). \qquad (6.65)$$

and

$$2DC_1\left[\frac{1}{RL}\cosh(R/L) - \frac{1}{R^2}\sinh(R/L)\right] = -\widehat{D}C_2'\left(\frac{1}{R\widehat{L}} + \frac{1}{R^2}\right)\exp(-R/\widehat{L}).$$

(6.66)

Solving this pair of equations for C_1 and C_2' and inserting the results into Eqs. (6.63) and (6.64) results in the solution:

$$\phi(r) = \left[1 - C\frac{R\,\sinh(r/L)}{r\,\sinh(R/L)}\right]\frac{s_0'''}{\Sigma_a}, \quad 0 \leq r < R, \qquad (6.67)$$

and

$$\phi(r) = (1 - C)\frac{s_o''' R}{\Sigma_a r}\exp\left[-(r - R)/\widehat{L}\right], \quad R < r \leq \infty, \tag{6.68}$$

where

$$C = \left[1 + \frac{D}{\widehat{D}}\frac{(R/L)\ \coth\ (R/L) - 1}{(R/\widehat{L}) + 1}\right]^{-1}. \tag{6.69}$$

6.6 Diffusion Approximation Validity

The question naturally arises as to the circumstances under which the diffusion equation, Eq. (6.12) provides a reasonable approximation. Figure 6.3 shows polar plots of the directions of neutron travel for three points in space. In each, the length of the arrows indicates the number of neutrons traveling in that direction. The plot of Fig. 6.3a indicates a beam of neutrons all traveling within a very narrow cone of directions. This is the situation we stipulated for the uncollided flux in Chapter 2 in order to define the cross section. The use of Eq. (6.12) in such situations would be inappropriate, leading to gross errors in the prediction of the spatial distribution of neutrons.

Diffusion theory is valid for the situations shown in Figs. 6.3b and 6.3c. In Fig. 6.3b neutrons travel in all directions. More are traveling to the right than to the left, but the distribution is not highly peaked. Use of the diffusion approximation in this situation is appropriate, for it will reasonably represent the flux, which decreases as we move from left to right, because the net diffusion of neutrons is in that direction.

Figure 6.3c represents an isotropic distribution of neutrons, such as would occur if there were no spatial variation in the neutron flux at all. In this case the use of the diffusion equation remains valid, but

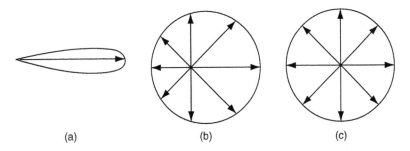

(a)   (b)   (c)

FIGURE 6.3 Polar distributions of neutrons. (a) Neutron beam, (b) neutrons diffusing to the right, (c) isotropic neutron distribution.

if the spatial variation vanishes, we would have $\vec{\nabla}\phi(\vec{r}) = 0$, and the leakage term would disappear from Eq. (6.12).

Generally, diffusion theory is valid if the neutron distribution is changing gradually in space. It tends to lead to significant errors for spatial distributions of neutrons near boundaries and at interfaces between materials with sharply different properties. In reactor lattices consisting of fuel, coolant, moderators, and/or other materials, we use diffusion theory to examine the global distribution of neutrons only after taking appropriate averages over the lattice cells, as discussed in Chapter 4.

Diffusion Length

We can gain a physical interpretation of the diffusion length by examining the mean square distance that a neutron diffuses between birth and absorption. To do this we assume that the neutron is born at $r = 0$ and then calculate the mean square distance weighted by the absorption rate $\Sigma_a\phi$:

$$\overline{r^2} = \frac{\int r^2 \Sigma_a \phi(r) dV}{\int \Sigma_a \phi(r) dV}. \tag{6.70}$$

Inserting Eq. (6.54) for the flux diffusing from a point source, and noting that $dV = 4\pi r^2 dr$, we have

$$\overline{r^2} = \frac{\int_0^\infty r^2 \Sigma_a \frac{S_p}{4\pi Dr} \exp(-r/L) 4\pi r^2 dr}{\int_0^\infty \Sigma_a \frac{S_p}{4\pi Dr} \exp(-r/L) 4\pi r^2 dr} = \frac{\int_0^\infty r^3 \exp(-r/L) dr}{\int_0^\infty r \exp(-r/L) dr}. \tag{6.71}$$

Evaluating the integrals then yields

$$\overline{r^2} = 6L^2, \tag{6.72}$$

or correspondingly

$$L = \frac{1}{\sqrt{6}} \sqrt{\overline{r^2}} = 0.408 \sqrt{\overline{r^2}}. \tag{6.73}$$

Thus the diffusion length is proportional to the root mean square (i.e., rms) distance diffused by a neutron between birth and absorption. During its life the neutron undergoes a number of scattering collisions, which change its direction of travel. Recall from Chapter 2 that $\lambda = 1/\Sigma$, the mean free path, is the average distance between such collisions. These quantities are easily related. Suppose we assume isotropic scattering. Then in Eq. (6.13) the diffusion coefficient is just $D = 1/(3\Sigma_t) = \lambda/3$. Next, we define the ratio of scattering to total cross section as

$$c = \Sigma_s/\Sigma_t. \tag{6.74}$$

Because $\Sigma_t = \Sigma_s + \Sigma_a$, it follows that the absorption cross section may be expressed as $\Sigma_a = (1 - c)\Sigma_t = (1 - c)/\lambda$. Finally, substituting these expressions for D and Σ_a into the definition of the diffusion length given by Eq. (6.15) we obtain

$$L = \lambda/\sqrt{3(1 - c)}. \tag{6.75}$$

Thus, for example, a material for which $c = 0.99$ would yield $L \approx 6\lambda$.

The relationship between mean free path and diffusion length is shown heuristically in Fig. 6.4. The dashed lines between scattering collisions average to a value of λ, while the length of the solid line is L. In general if c is less than about 0.7, the diffusion approximation predicated on Fick's approximation, given by Eq. (6.11), loses validity, and more advanced methods of neutron transport must be used.

Uncollided Flux Revisited

Further insight to the range of validity of the diffusion equation may be gained by comparing the uncollided flux from a point source to the total (i.e., uncollided + collided) flux attributable to the source. Recall from Eq. (2.9) that the uncollided flux from a point source is

$$\phi_u(r) = \frac{S_p}{4\pi r^2} \exp(-r/\lambda). \tag{6.76}$$

Comparing this expression to Eq. (6.54) reveals the following. Regardless of cross section the uncollided flux drops off as $1/r^2$ while the

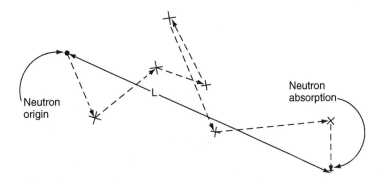

FIGURE 6.4   Diffusion length L between neutron birth and absorption.

total flux—uncollided plus collided—drops off only as $1/r$. Likewise, the mean free path appears in the exponential of the uncollided flux while the diffusion length—which is larger—appears in the exponential of the total flux. Thus the total flux will decay more slowly with distance.

A second point for comparison between uncollided and diffusing neutrons is the rms distances to first collision and to absorption, respectively. To obtain the rms distance traveled by uncollided neutrons we again consider a point source at the origin, but now we employ the first collision rate, $\Sigma_t \phi_u$, instead of the absorption rate:

$$\overline{r_u^2} = \frac{\int r^2 \Sigma_t \phi_u(r)dV}{\int \Sigma_t \phi_u(r)dV}. \tag{6.77}$$

Taking the uncollided flux from Eq. (6.76) and inserting it into definition for $\overline{r_u^2}$ yields

$$\overline{r_u^2} = \frac{\int_0^\infty r^2 \Sigma_t \dfrac{S_p}{4\pi r^2} \exp(-r/\lambda)4\pi r^2 dr}{\int_0^\infty \Sigma_t \dfrac{S_p}{4\pi r^2} \exp(-r/\lambda)4\pi r^2 dr} = \frac{\int_0^\infty r^2 \exp(-r/\lambda)dr}{\int_0^\infty \exp(-r/\lambda)dr}, \tag{6.78}$$

or evaluating the integrals:

$$\overline{r_u^2} = 2\lambda^2. \tag{6.79}$$

Or correspondingly, $\lambda = (1/\sqrt{2})\sqrt{\overline{r_u^2}} = 0.707\sqrt{\overline{r_u^2}}$, meaning that the mean free path is proportional to the rms distance traveled by a neutron before making its first collision. The ratio of the two mean square distances is

$$\frac{\overline{r^2}}{\overline{r_u^2}} = \frac{3L^2}{\lambda^2} = \frac{1}{1-c}. \tag{6.80}$$

As the distance from the source increases, the importance of the uncollided source rapidly diminishes provided that c, the ratio of scattering to total cross section, is sufficiently close to one. If it is not, then on average the neutrons will not make enough scattering collisions before absorption for the diffusion equation to be valid.

Examining angular distributions of neutrons further illuminates the distinction between uncollided and diffusing neutrons. The uncollided neutrons at a point some distance r from a point source all travel in a single direction, radially outward as shown schematically in Fig. 6.3a, while the diffusing neutrons will be dispersed in angle, more like Fig. 6.3b. For the diffusion equation to give reasonable accuracy, only a small fraction of the neutron population at a point can remain uncollided. Again, this condition typically exists only in situations where $c \equiv \Sigma_s / \Sigma_t$ is greater than about 0.7.

6.7 Multiplying Systems

We are now prepared to consider neutron diffusion in multiplying systems: those containing fissionable material. We frame the problems in spherical geometry, so that as we examine subcritical systems as they approach criticality the spatial effects are somewhat more realistic than in plane geometry. We defer treatment of the finite cylindrical form that power reactors invariably take to the following chapter; although spherical geometry can be treated using an ordinary differential equation, a partial differential equation must be solved for the finite cylinder.

Subcritical Assemblies

In a multiplying system, $\nu \Sigma_f > 0$ because fissionable material is present. If we limit our attention to a uniform system, with a uniform source s_o''', then the cross sections are space-independent constants. Dividing Eq. (6.12) by the diffusion coefficient D then yields

$$-\nabla^2 \phi(\vec{r}) + \frac{1}{L^2} \phi(\vec{r}) = \frac{1}{D} s_o''' + \frac{1}{L^2} k_\infty \phi(\vec{r}), \qquad (6.81)$$

where we again employ $k_\infty = \nu \Sigma_f / \Sigma_a$ and $L^2 = D / \Sigma_a$. Replacing ∇^2 by its one-dimensional form in spherical geometry reduces the diffusion equation to

$$-\frac{1}{r^2} \frac{d}{dr} r^2 \frac{d}{dr} \phi(r) + \frac{1}{L^2} (1 - k_\infty) \phi(r) = \frac{s_o'''}{D}. \qquad (6.82)$$

We look for a solution once more that is a superposition of general and particular solutions,

$$\phi(r) = \phi_g(r) + \phi_p(r). \qquad (6.83)$$

For a uniform source we hypothesize a constant for the particular solution. The derivative term thus vanishes from Eq. (6.82), and we obtain

$$\phi_p = \frac{s_0'''}{(1 - k_\infty)\Sigma_a}. \tag{6.84}$$

The general solution must satisfy

$$\frac{1}{r^2}\frac{d}{dr}r^2\frac{d}{dr}\phi_g(r) - \frac{1}{L^2}(1 - k_\infty)\phi_g(r) = 0. \tag{6.85}$$

With the same substitution as in Eq. (6.47),

$$\phi_g(r) = \frac{1}{r}\psi(r), \tag{6.86}$$

the equation simplifies to

$$\frac{d^2}{dr^2}\psi(r) - \frac{1}{L^2}(1 - k_\infty)\psi(r) = 0. \tag{6.87}$$

The form of the solution depends on whether $k_\infty < 1$ or $k_\infty > 1$. We first consider $k_\infty < 1$, and look for a solution of the form used earlier:

$$\psi(r) = C\exp(\kappa r). \tag{6.88}$$

Because

$$\frac{d^2}{dr^2}\psi(r) = C\kappa^2\exp(\kappa r), \tag{6.89}$$

Eq. (6.87) is satisfied if $\kappa^2 = \frac{1}{L^2}(1 - k_\infty)$, or equivalently

$$\kappa = \pm\frac{1}{L}\sqrt{1 - k_\infty}. \tag{6.90}$$

Thus there are two solutions, each of which may be multiplied by an arbitrary constant:

$$\psi(r) = C_1\exp(L^{-1}\sqrt{1 - k_\infty}\,r) + C_2\exp(-L^{-1}\sqrt{1 - k_\infty}\,r). \tag{6.91}$$

Inserting this expression into Eq. (6.86), and combining the result with Eqs. (6.83) and (6.84), we obtain for the flux

$$\phi(r) = \frac{C_1}{r}\exp(L^{-1}\sqrt{1 - k_\infty}\,r) + \frac{C_2}{r}\exp(-L^{-1}\sqrt{1 - k_\infty}\,r) + \frac{s_0'''}{(1 - k_\infty)\Sigma_a}. \tag{6.92}$$

We next apply the boundary conditions to determine C_1 and C_2. We can achieve condition that $\phi(0)$ must be finite only by requiring the two exponential terms to cancel exactly when $r=0$. Thus we take $C_2 = -C_1$. Then with the definition $\sinh(x) = \frac{1}{2}(e^x - e^{-x})$, we have

$$\phi(r) = \frac{2C_1}{r}\sinh(L^{-1}\sqrt{1-k_\infty}\,r) + \frac{s_o'''}{(1-k_\infty)\Sigma_a}. \tag{6.93}$$

With $\tilde{R}$ taken to be the extrapolated radius of the sphere, the other boundary condition is $\phi(\tilde{R}) = 0$, and from the above equation,

$$0 = \frac{2C_1}{\tilde{R}}\sinh(L^{-1}\sqrt{1-k_\infty}\,\tilde{R}) + \frac{s_o'''}{(1-k_\infty)\Sigma_a}. \tag{6.94}$$

Solving for C_1 and inserting the result into Eq. (6.93) yields

$$\phi(r) = \frac{s_o'''}{(1-k_\infty)\Sigma_a}\left[1 - \frac{\tilde{R}}{r}\frac{\sinh(L^{-1}\sqrt{1-k_\infty}\,r)}{\sinh(L^{-1}\sqrt{1-k_\infty}\,\tilde{R})}\right]. \tag{6.95}$$

For $k_\infty < 1$, the system must be subcritical, even if there were no neutrons leaking from the sphere. We next consider the system for $k_\infty > 1$, where criticality becomes possible. Once again we must have a general and a particular solution as in Eq. (6.83); Eq. (6.84) remains the particular solution, and Eqs. (6.85) through (6.87) remain applicable to the general solution. However, the general solution takes a different form. This is most easily seen by noting that for $k_\infty > 1$ the second term of Eq. (6.87) is now positive:

$$\frac{d^2}{dr^2}\psi(r) + \frac{1}{L^2}(k_\infty - 1)\psi(r) = 0. \tag{6.96}$$

For this differential equation we hypothesize a solution of the form

$$\psi(r) = C_1\sin(Br) + C_2\cos(Br). \tag{6.97}$$

Because

$$\frac{d^2}{dr^2}\psi(r) = -C_1B^2\sin(Br) - C_2B^2\cos(Br), \tag{6.98}$$

Eq. (6.96) is satisfied, provided we take

$$B = L^{-1}\sqrt{k_\infty - 1}. \tag{6.99}$$

Hence,

$$\psi(r) = C_1 \sin(L^{-1}\sqrt{k_\infty - 1}\, r) + C_2 \cos(L^{-1}\sqrt{k_\infty - 1}\, r). \qquad (6.100)$$

Analogous to the $k_\infty < 1$ case, we insert this expression into Eq. (6.86) and combine the result with Eqs. (6.83) and (6.84) to obtain the flux distribution:

$$\phi(r) = \frac{C_1}{r}\sin(L^{-1}\sqrt{k_\infty - 1}\, r) + \frac{C_2}{r}\cos(L^{-1}\sqrt{k_\infty - 1}\, r) - \frac{s_o'''}{(k_\infty - 1)\Sigma_a}. $$

$$(6.101)$$

We determine the constants by applying the same boundary conditions as in the $k_\infty < 1$ case: The condition at the origin is that $\phi(0)$ must be finite. The first term is finite since $\lim_{r \to 0} r^{-1} \sin(Br) = B$. Because $\cos(0) = 1$, however, the second term becomes infinite unless we set $C_2 = 0$. Consequently,

$$\phi(r) = \frac{C_1}{r}\sin(L^{-1}\sqrt{k_\infty - 1}\, r) - \frac{s_o'''}{(k_\infty - 1)\Sigma_a}. \qquad (6.102)$$

We determine C_1 by requiring this equation to meet the boundary condition $\phi(\tilde{R}) = 0$, yielding

$$\phi(r) = \frac{s_o'''}{(k_\infty - 1)\Sigma_a}\left[\frac{\tilde{R}}{r}\frac{\sin\left(L^{-1}\sqrt{k_\infty - 1}\, r\right)}{\sin\left(L^{-1}\sqrt{k_\infty - 1}\, \tilde{R}\right)} - 1\right]. \qquad (6.103)$$

Figure 6.5 shows the spatial distribution of $\phi(r)$ for increasing values of k_∞, first using Eq. (6.95) for $k_\infty < 1$ and then Eq. (6.103) for $k_\infty > 1$.

The Critical Reactor

Figure 6.5 indicates that the flux level increases with the value of k_∞. With increasing k_∞, the argument of the sine in the denominator of Eq. (6.103) becomes larger, until

$$L^{-1}\sqrt{k_\infty - 1}\, \tilde{R} = \pi, \qquad (6.104)$$

at which point the flux level becomes infinite, since $\sin(\pi) = 0$. At this point the sphere has become critical. As we saw in the preceding chapter subcritical reactors have time-independent solutions in the

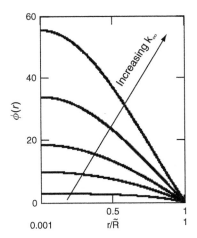

FIGURE 6.5   Flux distributions in subcritical spheres.

presence of a source, but critical reactors do not. Thus we should expect the solution given by Eq. (6.103) to become singular for a critical system.

Recall that for a finite reactor the criticality condition on the multiplication is just $k = P_{NL} k_\infty = 1$, where P_{NL} is the nonleakage probability. Squaring Eq. (6.104) and rearranging terms, we may write

$$1 = \frac{k_\infty}{1 + \left(\pi L/\tilde{R}\right)^2}. \tag{6.105}$$

Thus for the critical sphere, the nonleakage probability is

$$P_{NL} = \frac{1}{1 + \left(\pi L/\tilde{R}\right)^2}, \tag{6.106}$$

or, using the notation introduced in Chapter 5, $\Gamma = \left(\pi L/\tilde{R}\right)^2$. As expected the nonleakage probability increases with $\tilde{R}/L$, the extrapolated radius of the reactor measured in diffusion lengths.

Just as a subcritical reactor has a time-independent solution if and only if there is a source present, a critical reactor has a solution only if no source is present. For the sphere, this is equivalent to specifying that the general solution, Eq. (6.85), must be satisfied; it must also meet the same conditions: finite $\phi(0)$ and $\phi(\tilde{R}) = 0$; it must also guarantee that $\phi(r) > 0$ for $0 \le r < \tilde{R}$. In fact, if we just set $s_o''' = 0$ in Eq. (6.102), we obtain

$$\phi(r) = \frac{C_1}{r} \sin(L^{-1} \sqrt{k_\infty - 1}\, r). \tag{6.107}$$

The condition $\phi(\tilde{R}) = 0$, along with the requirement that $\phi(r) > 0$, then yields Eq. (6.104), the criticality condition. Note that the solution is not unique; that is, C_1 may take on any nonnegative value. In Chapter 7 we will show that C_1 is proportional to the power at which the reactor operates.

In earlier texts, the term $B_m = L^{-1}\sqrt{k_\infty - 1}$ that first appeared (without the subscript) in Eq. (6.99) is referred to as the material buckling; it depends only on material properties. Likewise, for a sphere, $B_g = \pi/R$ is referred to as the geometric buckling; it depends only on the sphere's size. With this terminology the criticality condition of Eq. (6.105) may also be stated as the material and geometric buckling being equal: $B_g = B_m$. The term buckling derived from the notion that the flux had more concave downward or "buckled" curvature in a small reactor than in a large one. Thus if a reactor is smaller than the critical size for a given material, $B_g > B_m$, and the reactor is subcritical. If it is larger than the critical size, $B_g < B_m$, and the reactor is supercritical. For reactors of shapes other than spheres the geometrical buckling takes the form $B_g = C/R$, where the coefficient C is determined by the shape of the reactor and R is a characteristic dimension. Generally, the multiplication of a uniform reactor of any shape and size is given by $k = k_\infty P_{NL}$, with the nonleakage probability written as

$$P_{NL} = \frac{1}{1 + L^2 B^2}, \qquad (6.108)$$

where the subscript is dropped from B, the geometric buckling. The following chapter includes the derivation of the geometric buckling for the cylindrical form of a power reactor core.

Bibliography

Duderstadt, James J., and Louis J. Hamilton, *Nuclear Reactor Analysis*, Wiley, NY, 1976

Henry, Allen F., *Nuclear-Reactor Analysis*, MIT Press, Cambridge, MA, 1975.

Knief, Ronald A., *Nuclear Energy Technology: Theory and Practice of Commercial Nuclear Power*, McGraw-Hill, NY, 1981.

Lamarsh, John R., *Introduction to Nuclear Reactor Theory*, Addison-Wesley, Reading, MA, 1972.

Lewis, E. E., and W. F. Miller Jr., *Computational Methods of Neutron Transport*, Wiley, NY, 1984.

Meghreblian, R. V., and D. K. Holmes, *Reactor Analysis*, McGraw-Hill, NY, 1960.

Ott, Karl O., and Winfred A. Bezella, *Introductory Nuclear Reactor Statics*, 2nd ed., American Nuclear Society, 1989.
Stacey, Weston M., *Nuclear Reactor Physics*, Wiley, NY, 2001.

Problems

6.1. Consider in plane geometry a slab of nonfissionable material with properties D and L and a thickness of a. Assume s'' neutrons/cm^2/s enter the slab from the left. What fraction of the neutrons will

   a. Penetrate the slab?
   b. Be reflected back from the left-hand surface?
   c. Be absorbed in the slab?

6.2. Consider an infinite slab of nonfissionable material occupying the region $-a \leq x \leq a$. It contains a uniform source s_o''' and has properties D_1 and Σ_{a1}. A second nonfissionable material, with D_2 and Σ_{a2}, contains no source and occupies the remaining regions $-\infty \leq x \leq -a$ and $a \leq x \leq \infty$. Find the flux distribution for $0 \leq x \leq \infty$.

6.3. Determine the fraction of neutrons that penetrate a 1-m-thick slab of graphite for which $D = 0.84$ cm and $\Sigma_a = 2.1 \times 10^{-4}$ cm^{-1}. Evaluate the albedo of the slab.

6.4. In plane geometry, thermal neutrons enter a nonmultiplying slab of infinite thickness from the left. The properties of the composite slab are D_1 and Σ_{a1} for $0 \leq x \leq a$ and D_2 and Σ_{a2} for $a \leq x \leq \infty$.

   a. Show that with $\Upsilon_i \equiv D_i/L_i$ the albedo may be expressed as

$$\alpha = \frac{(1 - 2\Upsilon_1) + \dfrac{\Upsilon_2 - \Upsilon_1}{\Upsilon_2 + \Upsilon_1}(1 + 2\Upsilon_1)\exp(-2a/L_1)}{(1 + 2\Upsilon_1) + \dfrac{\Upsilon_2 - \Upsilon_1}{\Upsilon_2 + \Upsilon_1}(1 - 2\Upsilon_1)\exp(-2a/L_1)}.$$

   b. Evaluate the albedo for $a = 10$ cm when material 1 is graphite and material 2 is water.
   c. Repeat part b when materials 1 and 2 are switched.

6.5. Note that if $2D/L > 1$, Eq. (6.41) yields a negative albedo! Explain in terms of cross sections why diffusion theory is not valid under these circumstances.

6.6. Neutrons impinge uniformly over the surface of a sphere made of graphite that has a diameter of 1.0 m. For the graphite $D = 0.84$ cm and $\Sigma_a = 2.1 \times 10^{-4}$ cm^{-1}.

    a. Determine the albedo of the graphite sphere.

    b. Determine the fraction of the impinging neutrons that are absorbed in the sphere.

6.7. Verify Eq. (6.55).

6.8. In the text the spherical geometry condition $0 < \phi(0) < \infty$ is employed to reduce Eq. (6.60) to Eq. (6.63). Show that the condition $\lim_{r\to 0} 4\pi r^2 J_r(r) = 0$ produces the same result.

6.9. The point source of strength s_p is located at the center of a sphere of a nonmultiplying medium with properties D and Σ_a, and an extrapolated radius $\tilde{R}$, in an infinite vacuum.

    a. Find the flux distribution in the sphere.

    b. Determine the fraction of neutrons that escape from the sphere without being absorbed if $\tilde{R} = L$.

6.10. A thin spherical shell of radius R emits s''_{pl} neutrons/cm²/s in an infinite nonmultiplying medium with properties D and Σ_a.

    a. Determine the flux $\phi(r)$ for $0 \le r \le \infty$.

    b. Determine the flux ratio $\phi(0)/\phi(R)$.

6.11.*Show that the ratio of uncollided to total flux from a point source in an infinite medium is

$$\frac{\phi_u(r)}{\phi(r)} = \sqrt{1/3(1-c)}\,\frac{1}{(r/L)}\,exp\,[-\alpha(r/L)]$$

where $\alpha = [3(1-c)]^{-1/2}-1$. Then plot the curve for $1/2 < r/L < 3$ using thermal neutron cross sections for the following:

    a. Water.

    b. Heavy water.

    c. Graphite.

    d. A one-to-one volume mixture of natural uranium and water.

6.12. A semi-infinite multiplying medium having the properties D, Σ_a, and $k_\infty < 1$ occupies the space $0 \le x \le \infty$, while $-\infty \le x < 0$ is a vacuum. A source embedded in the medium emits neutrons at a rate of $S'''_o \exp(-\alpha x)$ neutrons/cm³ s⁻¹. Neglecting the extrapolation distance, show that the distribution of neutrons is

$$\phi(x)=[\alpha^2-(1-k_\infty)/L^2]^{-1}\frac{S'''_o}{D}\left[\exp(-\sqrt{1-k_\infty}x/L)-\exp(-\alpha x)\right],$$
$$0\le x\le\infty.$$

6.13. Verify Eqs. (6.67) through (6.69).

6.14. Show that Eqs. (6.95) and (6.103) agree in the limit of $k_\infty \to 1$.

6.15. Suppose that the material in problem 6.9 is fissionable with $k_\infty < 1$. Find the flux distribution in the sphere.

6.16. Suppose the material in problem 6.9 is fissionable with $k_\infty > 1$:

a. Find the flux distribution in the sphere.
b. Show that the criticality condition is the same as Eq. (6.105).

6.17. Equations (6.95) and (6.103) give the flux distributions for a subcritical sphere with a uniform source for $k_\infty < 1$ and $k_\infty > 1$, respectively. Find the equivalent expression for $k_\infty = 1$.

6.18. Using Eqs. (6.95) and (6.103),

a. Find expressions for the flux $\phi(0)$ at the center of the subcritical sphere.
b. Using your results from part a make a plot of $\phi(0)$ for $0 \le k_\infty < 1.154$ with $\tilde{R}/L = 8$.
c. Using your results from part a make a plot of $\phi(0)$ for $0 < \tilde{R}/L < 8$, with $k_\infty = 1.154$.
d. Compare the two curves and discuss their significance. (Normalize plots to S_o''' / Σ_a)

CHAPTER 7

Neutron Distributions in Reactors

7.1   Introduction

Chapter 6 concluded with the approach to criticality of a spherical system, chosen for the mathematical simplicity of its one-dimensional geometry. This chapter deals with the spatial distributions of neutrons within the finite cylindrical volumes that correspond to the cores of power reactors. We begin by reformulating the diffusion equation in an eigenvalue form that yields both the multiplication and the flux distribution from a time-independent solution. We then find the criticality equation and flux distribution for a bare uniform reactor and relate it to the reactor power. Uniform, remember, means only that the lattice cells are identical, for in our approximate treatment we assume that the cross sections are averaged over energy—allowing an energy-independent treatment—and over the cross-sectional area of the lattice cell. The next section provides a more detailed treatment of the neutron nonleakage probability and the effects that neutron slowing down and diffusion have on it. Following completion of our treatment of the bare cylindrical reactor, we examine reactors that include reflector regions to improve the neutron economy. We conclude the chapter with an examination of the effects control poisons—first of a single control rod and then of a bank of control rods—on reactor multiplication and flux distribution.

7.2   The Time-Independent Diffusion Equation

With the source set to zero, the steady state diffusion equation, given by Eq. (6.12), for the flux distribution within a fissionable system becomes

$$\vec{\nabla} \cdot D\vec{\nabla}\phi + \nu\Sigma_f\phi - \Sigma_a\phi = 0, \tag{7.1}$$

167

where for brevity we have deleted the $(\vec{r})$ that denotes spatial dependence. This equation has a positive flux solution within a reactor only if it is exactly critical. Otherwise, the neutron population will vary with time, and the kinetics equations of Chapter 5 are needed to describe the system's behavior. Often, however, the problem at hand involves searching for the critical state by varying the reactor's geometry (its radius or height, for example) or its material composition (for example, its fuel enrichment or its fuel to moderator ratio). We do such searches iteratively by hypothesizing a reactor of a particular size, shape, and composition and determining how far from critical the system is and what the spatial distribution of neutrons within it will be. This we accomplish without carrying out detailed time-dependent calculations through the following artifice.

Suppose that we could vary the average number of neutrons per fission, ν, by a ratio ν_o/ν. Equation (7.1) then becomes

$$\vec{\nabla} \cdot D\vec{\nabla}\phi + (\nu_o/\nu)\nu\Sigma_f\phi - \Sigma_a\phi = 0. \tag{7.2}$$

Now suppose that ν_o is the number of neutrons per fission that we would require to make the reactor configuration exactly critical (i.e., it would yield a multiplication of $k = 1$), while ν is the number of neutrons actually produced per fission. Since k is always proportional to the number of neutrons per fission, we have $\nu_o/\nu = 1/k$, and we may write Eq. (7.2) as

$$\vec{\nabla} \cdot D\vec{\nabla}\phi + \frac{1}{k}\nu\Sigma_f\phi - \Sigma_a\phi = 0. \tag{7.3}$$

If the number of neutrons per fission, ν_o, needed to make the reactor exactly critical is larger than the actual value, ν, then the reactor is subcritical, and k is less than one. Conversely, the reactor is supercritical if ν_o is smaller than ν, and k is greater than one.

By solving Eq. (7.3) for k, the foregoing technique converts the problem from trying to find a combination of cross sections and dimensions for which a solution of Eq. (7.1) exists to specifying a set of cross sections and dimensions and determining how far from critical the configuration is. Equation (7.3) has the form of an eigenvalue problem, where k is the eigenvalue and ϕ is the eigenfunction. In general there will be many—in some cases an infinite number—of eigenvalues and eigenfunctions that solve Eq. (7.3) and meet the appropriate boundary conditions. We are only interested, however, in the physically meaningful solution for which the flux is positive everywhere within the reactor volume. This solution may be shown to correspond to the largest eigenvalue, which is the multiplication; we refer to the corresponding eigenfunction ϕ, which is positive everywhere within the reactor, as the fundamental mode solution.

7.3   Uniform Reactors

We begin by analyzing a reactor that is uniform throughout. Accordingly, none of the cross sections vary in space. Thus D is a constant, and using our earlier definitions for $k_\infty = \nu\Sigma_f/\Sigma_a$ and $L^2 = D/\Sigma_a$ we may write Eq. (7.3) as

$$\nabla^2\phi + \frac{k_\infty/k - 1}{L^2}\phi = 0, \tag{7.4}$$

or equivalently

$$-\frac{\nabla^2\phi}{\phi} = \frac{k_\infty/k - 1}{L^2}. \tag{7.5}$$

Since the right side of this equation is independent of the spatial variables, so must the left be the same; both sides of the equation must be equal to the same constant for it to be satisfied. Specifying that constant as $\nabla^2\phi/\phi = -B^2$, we obtain for the multiplication

$$k = \frac{k_\infty}{1 + L^2 B^2}. \tag{7.6}$$

To determine B, referred to as the geometric buckling or simply as the buckling, we must solve

$$\nabla^2\phi + B^2\phi = 0, \tag{7.7}$$

which is a Helmholtz equation. The solution, moreover, must meet the condition $0 < \phi < \infty$ within the reactor, and well as satisfying the boundary conditions at its surfaces.

We have already encountered such a problem. Recall that we concluded Chapter 6 by going from subcritical to critical for an idealized spherical reactor. We obtained an expression, Eq. (6.105), which is identical in form to Eq. (7.6), but with k set equal to one and with the buckling for the sphere given by $B = \pi/R$. And, the nonleakage probability of Eq. (6.106) takes the form

$$P_{NL} = \frac{1}{1 + L^2 B^2}. \tag{7.8}$$

Equations (7.6) through (7.8) are valid for uniform reactors of all shapes. In what follows, we analyze a cylindrical reactor of finite length to determine the multiplication and flux distribution in terms of reactor's dimensions and material properties.

Finite Cylindrical Core

For a uniform cylindrical reactor with an extrapolated radius $\tilde{R}$ and extrapolated height $\tilde{H}$, as shown in Fig. 7.1, Eq. (7.7) is a partial differential equation. It may be expressed in terms of the radial and axial coordinates r and z. With the r–z geometry form of the ∇^2 operator given in Appendix A, we have

$$\frac{1}{r}\frac{\partial}{\partial r}r\frac{d}{dr}\phi + \frac{\partial^2}{\partial z^2}\phi + B^2\phi = 0, \qquad (7.9)$$

subject to the conditions

$$0 < \phi(r,z) < \infty, \quad 0 \leq r \leq \tilde{R}, \quad -\tilde{H}/2 \leq z \leq \tilde{H}/2. \qquad (7.10)$$

We look for a solution by separating variables, that is, by separating $\phi(r,z)$ into a product of functions of r and of z:

$$\phi(r,z) = \psi(r)\chi(z). \qquad (7.11)$$

Inserting this expression into Eq. (7.9) and dividing by $\psi\chi$, we obtain

$$\frac{1}{\psi r}\frac{d}{dr}r\frac{d}{dr}\psi + \frac{1}{\chi}\frac{d^2}{dz^2}\chi + B^2 = 0. \qquad (7.12)$$

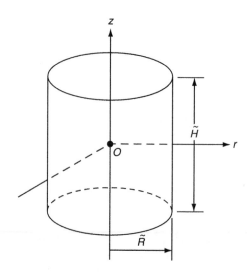

FIGURE 7.1   Cylindrical reactor core.

Because the first term depends only on r, and the second only on z, both terms must be constants for the equation to have a solution. Suppose we take the constants to be $-B_r^2$ and $-B_z^2$. The solution then takes the form

$$B_r^2 + B_z^2 = B^2, \qquad (7.13)$$

where B_r^2 and B_z^2 must satisfy the differential equations

$$\frac{1}{r}\frac{d}{dr}r\frac{d}{dr}\psi + B_r^2\psi = 0, \quad 0 \leq r \leq \tilde{R}, \qquad (7.14)$$

and

$$\frac{d^2}{dz^2}\chi + B_z^2\chi = 0, \quad -\tilde{H}/2 \leq z \leq \tilde{H}/2. \qquad (7.15)$$

Thus we have reduced Eq. (7.9), a partial differential equation, to two ordinary differential equations in r and z, respectively. Substituting

$$\chi(z) = C_1 \sin(B_z z) + C_2 \cos(B_z z), \quad -\tilde{H}/2 \leq z \leq \tilde{H}/2, \qquad (7.16)$$

into Eq. (7.15), we may show that it is a solution, where C_1 and C_2 are arbitrary constants to be determined from the boundary conditions. The boundary conditions at the ends of the cylinder are $\chi(\pm\tilde{H}/2) = 0$. Because the equation and its boundary conditions are symmetric in z about its mid plane, the solution must be symmetric: $\chi(z) = \chi(-z)$. Thus it follows that $C_1 = 0$, because $\sin(B_z z) = -\sin(-B_z z)$. Satisfying the axial boundary conditions then translates to the requirement $\cos(\pm B_z\tilde{H}/2) = 0$. This condition is met provided $B_z\tilde{H}/2 = \pi/2, 3\pi/2, 5\pi/2, \cdots$. Only for the $\pi/2$ root, however, will χ and hence the flux be positive everywhere in the core. We therefore take

$$B_z = \pi/\tilde{H}. \qquad (7.17)$$

For the radial direction, the solution of Eq. (7.14) takes the form of the less familiar Bessel functions, which are plotted in Fig. 7.2 and discussed further in Appendix B:

$$\psi(r) = C_1'J_0(B_r r) + C_2'Y_0(B_r r), \quad 0 \leq r \leq \tilde{R}, \qquad (7.18)$$

where C_1' and C_2' are arbitrary constants. We again have two arbitrary constants to determine. Noting from Fig. 7.2 that $Y_0(0) \to -\infty$, we take $C_2' = 0$; otherwise, the flux would become infinite along the centerline of the reactor. From the graph of $J_0(x)$, we see that in order for the flux

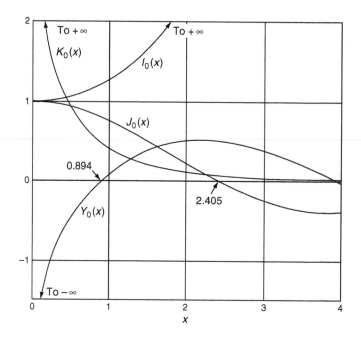

FIGURE 7.2   Ordinary and modified Bessel functions of order zero.

to vanish at the extrapolated radius $\tilde{R}$, but remain positive for all smaller values of r, we must have $B_r \tilde{R} = 2.405$, or correspondingly

$$B_r = 2.405/\tilde{R}. \tag{7.19}$$

We next combine the foregoing results to obtain the buckling and flux distribution for a cylindrical reactor. For the buckling we insert Eqs. (7.17) and (7.19) into Eq. (7.13):

$$B^2 = (2.405/\tilde{R})^2 + (\pi/\tilde{H})^2. \tag{7.20}$$

The flux distribution results from inserting Eqs. (7.16) and (7.18) into Eq. (7.11), subject to the restrictions $C_1 = 0$ and $C_2' = 0$:

$$\phi(r, z) = CJ_0(2.405r/\tilde{R}) \cos(\pi z/\tilde{H}), \tag{7.21}$$

where $C = C_1' C_2$.

Reactor Power

One arbitrary constant, C, remains. It is proportional to the power of the reactor. The amount of recoverable energy per fission varies slightly between fissionable isotopes. A reasonable average is

$\gamma = 3.1 \times 10^{-11}$ J/fission (i.e., watt-s/fission). Then because $\Sigma_f \phi$ is the number of fissions/cm^3/s the reactor power is just $\gamma \Sigma_f \phi$ integrated over the volume of the reactor:

$$P = \gamma \int \Sigma_f \phi dV, \qquad (7.22)$$

which in cylindrical geometry reduces to

$$P = \gamma 2\pi \int_0^{\tilde{R}} \int_{-\tilde{H}/2}^{\tilde{H}/2} \Sigma_f \phi dz r dr. \qquad (7.23)$$

Inserting Eq. (7.21) then yields

$$P = \gamma \Sigma_f 2\pi C \int_0^{\tilde{R}} J_0(2.405r/\tilde{R}) r dr \int_{-\tilde{H}/2}^{\tilde{H}/2} \cos(\pi z/\tilde{H}) dz. \qquad (7.24)$$

We may separate this expression into more convenient radial and axial contributions by multiplying numerator and denominator by the volume $V = \pi R^2 H$:

$$P = \gamma \Sigma_f VC \left[\frac{2}{\tilde{R}^2} \int_0^{\tilde{R}} J_0(2.405r/\tilde{R}) r dr \right] \left[\frac{1}{\tilde{H}} \int_{-\tilde{H}/2}^{\tilde{H}/2} \cos(\pi z/\tilde{H}) dz \right]. \qquad (7.25)$$

Changing variables $\xi = 2.405r/\tilde{R}$ and using the Bessel function identities found in Appendix B allows us to eliminate the radial integral:

$$\frac{2}{\tilde{R}^2} \int_0^{\tilde{R}} J_0(2.405r/\tilde{R}) r dr = \frac{2}{2.405^2} \int_0^{2.405} J_0(\xi) \xi d\xi = \frac{2}{2.405} J_1(2.405) \qquad (7.26)$$

and, likewise, substituting $\varsigma = \pi z/H$ yields

$$\frac{1}{\tilde{H}} \int_{-\tilde{H}/2}^{\tilde{H}/2} \cos(\pi z/\tilde{H}) dz = \frac{1}{\pi} \int_{-\pi/2}^{\pi/2} \cos(\varsigma) d\varsigma = \frac{2}{\pi} \qquad (7.27)$$

for the axial component. Noting that $J_1(2.405) = 0.519$, we combine the three preceding equations to obtain

$$P = \gamma \Sigma_f VC \frac{2J_1(2.405)}{2.405} \frac{2}{\pi} = 0.275 \gamma \Sigma_f VC, \qquad (7.28)$$

or

$$C = 3.63 \frac{P}{\gamma \Sigma_f V}. \qquad (7.29)$$

Finally, inserting C into Eq. (7.21) expresses the flux distribution in terms of the reactor power:

$$\phi(r, z) = 3.63 \frac{P}{\gamma \Sigma_f V} J_0(2.405r/R) \cos(\pi z/H). \qquad (7.30)$$

7.4   Neutron Leakage

In general, the cross sections and diffusion coefficients that we use in the diffusion approximation are averaged over the entire energy spectrum of neutrons as well as over the lattice cell compositions as discussed in Chapter 4. The diffusion length measures the distance that neutrons travel between birth and death. In thermal reactor calculations, however, the diffusion coefficient and cross sections used are often those averaged over only the thermal neutron energy spectrum. Uncorrected, such calculations neglect the distance that neutrons diffuse while slowing down to thermal energies. In some systems, particularly those with light water moderators, such neglect leads to significant error. Treating the neutron migration during both slowing down and thermal diffusion more rigorously requires dividing the neutron spectrum into two or more energy groups, and devising a diffusion equation for each. For thermal reactors, however, dividing the flux into just two groups—fast and thermal—is often adequate, particularly when considering the lattice physics in terms of the four factor formula developed in Chapter 4. Figure 7.3 illustrates the same neutron cycle as Fig. 4.5, but with neutron diffusion added for both fast neutrons, as they slow down through the intermediate or resonance regions, as well as for thermal neutrons.

Two Group Approximation

To model a reactor with two group theory, we begin by defining the fast and thermal flux as ϕ_1 and ϕ_2. To obtain the source of neutrons for fast diffusion, we note that the number of thermal neutrons absorbed is $\Sigma_a \phi_2/\text{cm}^3/\text{s}$ and that of these $f\Sigma_a \phi_2/\text{cm}^3/\text{s}$ are absorbed in fuel. We next multiply by η_T to obtain the number of fast neutrons produced from thermal fission. However, we also include $1/k$ to take

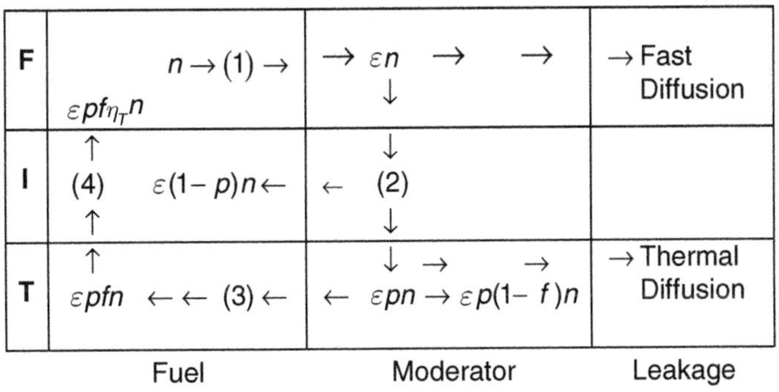

F	$n \to (1) \to$ $\varepsilon pf\eta_T n$	$\to \varepsilon n \to \qquad \to$ $\downarrow$	$\to$ Fast Diffusion
I	$\uparrow$ (4) $\quad \varepsilon(1-p)n \leftarrow$ $\uparrow$	$\downarrow$ $\leftarrow \quad$ (2) $\downarrow$	
T	$\uparrow$ $\varepsilon pfn \quad \leftarrow\leftarrow (3) \leftarrow$	$\downarrow \to \qquad \to$ $\leftarrow \varepsilon pn \to \varepsilon p(1-f)n$	$\to$ Thermal Diffusion
	Fuel	Moderator	Leakage

FIGURE 7.3 Four factor formula with leakage for a thermal reactor neutron cycle.

into account our varying of the neutrons per fission as in Eq. (7.3) in order to obtain a steady state neutron distribution even though the reactor is not exactly critical. Finally, we multiply by ε, the fast fission factor, and obtain $(1/k)\varepsilon\eta_T f \Sigma_a \phi_2$ as the number of fast neutrons produced/cm^3/s. The diffusion equation for fast neutrons is

$$-\vec{\nabla} \cdot D_1 \vec{\nabla}\phi_1 + \Sigma_r \phi_1 = \frac{1}{k}\varepsilon\eta_T f \Sigma_a \phi_2. \tag{7.31}$$

The first term on the left is the fast leakage. It has the same form as Eq. (7.1) except

$$D_1 = 1/3\Sigma_{tr1} \tag{7.32}$$

is the diffusion coefficient for fast neutrons. The second term $\Sigma_r \phi_1$ accounts for the neutrons removed from the fast group by slowing down. In paragraphs that follow we will discuss the calculation of D_1 and the removal cross section Σ_r.

The source of neutrons for thermal diffusion is the same as those removed from the fast group, multiplied by the resonance escape probability p to account for those lost to capture in the fuel resonance capture cross sections: $p\Sigma_r \phi_1$/cm^3/s; Thus the thermal diffusion equation is

$$-\vec{\nabla} \cdot D_2 \vec{\nabla}\phi_2 + \Sigma_a \phi_2 = p\Sigma_r \phi_1, \tag{7.33}$$

where the first term accounts for thermal neutron leakage, and the second for thermal absorption, where

$$D_2 = 1/3\Sigma_{tr2} \tag{7.34}$$

and Σ_a are, respectively, the thermal diffusion coefficient and absorption cross section.

For a uniform region the diffusion coefficients are space independent and may be pulled outside the divergence operator. Dividing the two diffusion equations by Σ_r and Σ_a, respectively, and defining fast and thermal diffusion lengths by

$$L_1 = \sqrt{D_1/\Sigma_r} \qquad (7.35)$$

and

$$L_2 = \sqrt{D_2/\Sigma_a}, \qquad (7.36)$$

we obtain

$$-L_1^2 \nabla^2 \phi_1 + \phi_1 = \frac{1}{k} \varepsilon \eta_T f \frac{\Sigma_a}{\Sigma_r} \phi_2 \qquad (7.37)$$

and

$$-L_2^2 \nabla^2 \phi_2 + \phi_2 = p \frac{\Sigma_r}{\Sigma_a} \phi_1. \qquad (7.38)$$

Next consider a uniform reactor with zero flux boundary conditions on all its extrapolated outer surfaces. Similar to the technique applied to the one group model in the preceding section, we may represent the spatial dependence of the flux with Helmholtz equations:

$$\nabla^2 \phi_1 + B^2 \phi_1 = 0 \qquad (7.39)$$

and

$$\nabla^2 \phi_2 + B^2 \phi_2 = 0. \qquad (7.40)$$

Using these equations to replace the ∇^2 terms in Eqs. (7.37) and (7.38) we obtain

$$\phi_1 = \frac{1}{1 + L_1^2 B^2} \frac{1}{k} \varepsilon \eta_T f \frac{\Sigma_a}{\Sigma_r} \phi_2 \qquad (7.41)$$

and

$$\phi_2 = \frac{1}{1 + L_2^2 B^2} p \frac{\Sigma_r}{\Sigma_a} \phi_1. \qquad (7.42)$$

Finally, combining equations we have

$$k = \frac{1}{1 + L_1^2 B^2} \frac{1}{1 + L_2^2 B^2} k_\infty, \qquad (7.43)$$

where the infinite medium multiplication is given by the four factor formula: $k_\infty = p\varepsilon f \eta_T$. Note that we may now write—as in Eq. (3.2)— $k = k_\infty P_{NL}$, provided we replace the nonleakage probability given by Eq. (7.8) with

$$P_{NL} = \frac{1}{1 + L_1^2 B^2} \frac{1}{1 + L_2^2 B^2}. \qquad (7.44)$$

The definition of the thermal diffusion length is straightforward, with both the transport and absorption cross sections appearing in Eqs. (7.34) and (7.36) averaged over the thermal neutron spectrum. However, the diffusion length for fast neutrons requires more careful scrutiny. The Fermi age is the most frequently employed approximation for L_1^2. It is defined as

$$\tau = \int_{E_2}^{E_1} \frac{D(E)}{\xi \Sigma_s(E) E} dE, \qquad (7.45)$$

where the integral extends from fission energy neutrons down to thermal neutrons, typically from $E_1 = 2.0\,\text{MeV}$ to $E_2 = 0.0253\,\text{eV}$. Note that the slowing down power, $\xi \Sigma_s$, introduced in Chapter 2, appears in the denominator. Thus taking

$$L_1^2 = \tau, \qquad (7.46)$$

we see that a strong moderator results in a small diffusion length for fast neutrons. We may gain some further insight with the following simplified model. Over the energy range included in Eq. (7.45) we approximate the diffusion coefficient and scattering cross section as energy independent. Equation (7.46) thus reduces to

$$L_1^2 \approx \frac{D_1}{\xi \Sigma_s} \ln(E_1/E_2). \qquad (7.47)$$

Combining this result with Eq. (7.35) allows us to write the removal cross section as $\Sigma_r = \xi \Sigma_s / \ln(E_1/E_2)$. Next, recall from Eq. (2.59) that the estimated number of elastic scattering collisions required to slow a neutron down from E_1 to E_2 is $n \approx (1/\xi) \ln(E_1/E_2)$. Hence we may approximate the removal cross section as $\Sigma_r \approx \Sigma_s/n$. Because

$L_1 \approx \sqrt{nD_1/\Sigma_s}$, the smaller the number of collisions required to slow the fission neutrons to thermal energies, the shorter the distance that fast neutrons will diffuse before being slowed to thermal energies.

Migration Length

Retaining the one group theory developed in earlier sections for thermal as well as fast reactors simplifies further analysis considerably. For large reactors, we can accomplish this without a substantial loss of accuracy simply by replacing the diffusion length L by the migration length M in the one group equations. To define the migration length we multiply the nonleakage contributions from the two groups, found in Eq. (7.44), together and obtain

$$P_{NL} = \frac{1}{1 + (L_1^2 + L_2^2)B^2 + L_1^2 L_2^2 B^4}. \tag{7.48}$$

Since B^2 is small for large reactors, a reasonable approximation is to drop the B^4 term from the denominator. We may then write

$$P_{NL} = \frac{1}{1 + M^2 B^2}, \tag{7.49}$$

where the migration area is defined as

$$M^2 = L_1^2 + L_2^2, \tag{7.50}$$

and correspondingly the migration length is

$$M = \sqrt{L_1^2 + L_2^2}. \tag{7.51}$$

Table 7.1 lists the values of L_1 $(=\sqrt{\tau})$, L_2, and M for the three most common moderators and also representative values for power reactors using each of them. The table indicates that the largest correction to the thermal diffusion length is for water-moderated systems. This comes about primarily because hydrogen's large thermal absorption cross section causes L_2 to be quite small compared to reactors utilizing other moderators. To lesser extent, the hydrogen's decreased scattering cross section over the energy range where fission neutrons are born, pictured in Fig. 2.3a, increases the value of L_1 since Σ_s appears in the denominator of Eq. (7.45). For fast reactors the diffusion and migration length are considered one and the same. To complete Table 7.1, a representative value for a sodium-cooled fast reactor (SFR) is $M = 19.2$ cm and for a gas-cooled fast reactor (GCFR), $M = 25.5$ cm.

TABLE 7.1
Representative Diffusion Properties for Moderators and Thermal Reactors

Type	Description	$L_1 = \sqrt{\tau}$ *(cm)* *Fast Diffusion Length*	L_2 *(cm)* *Thermal Diffusion Length*	*M (cm)* *Migration Length*
H_2O	Light Water	5.10	2.85	5.84
PWR	Pressurized-H_2O Reactor	7.36	1.96	7.62
BWR	Boiling-H_2O reactor	7.16	1.97	7.43
D_2O	Heavy water	11.5	173	174
PHWR	CANDU-D_2O reactor	11.6	15.6	19.4
C	Graphite	19.5	59.0	62.0
HTGR	Graphite-moderated He-cooled reactor	17.1	10.6	20.2

Source: Data courtesy of W. S. Yang, Argonne National Laboratory.

Leakage and Design

To examine the relationship between neutron leakage and power reactor design, we begin by combining Eq. (7.49) with $k = P_{NL}k_\infty$ to obtain

$$k = \frac{1}{1 + M^2 B^2} k_\infty. \qquad (7.52)$$

Suppose we make this expression more explicit by assuming that the reactor is a cylinder with a height to diameter ratio of one. Then with $\tilde{H} = 2\tilde{R}$ the buckling given by Eq. (7.20) becomes $B^2 = 33.0/\tilde{H}^2$, and we may write

$$k = \frac{1}{1 + 33.0(M/\tilde{H})^2} k_\infty. \qquad (7.53)$$

Thus the primary determinant of neutron leakage is $\tilde{H}/M$, the characteristic dimension of the reactor, measured in migration lengths.

Very roughly speaking the reactor design process may be characterized as follows. Ordinarily, the power P that the reactor must be capable of producing has been set before the design commences. The designers first fix the structure of the core lattice, selecting the fuel, moderator, coolant, and other materials, their volume ratios, and their geometrical configuration (i.e., fuel radius, lattice pitch, etc.).

In doing so they attempt to select lattice parameters such that (a) for a given fuel enrichment the value of k_∞ is near optimum and (b) the power per unit volume (i.e., P''', the power density) that can be transported from fuel to coolant outlet is maximized.

Because the core materials and lattice parameters largely determine the value of the migration length—which is, however, only very weakly dependent on the fuel enrichment—at this point the value of M is pretty much fixed. The core lattice design taken together with maximum to average flux ratio determines $\bar{P}'''$, the achievable core averaged power density. Then with P and $\bar{P}'''$ determined, the relationship $P = V\bar{P}'''$ determines the core volume. For a cylindrical reactor with a height to diameter ratio of one, the volume is $V = \pi \tilde{H}^3/4$ and consequently $\tilde{H} = \sqrt[3]{4V/\pi} = \sqrt[3]{4P/\pi\bar{P}'''}$. Thus the core lattice design determines M, and the achievable $\bar{P}'''$, whereas a reactor's core volume increases linearly with the required output P of the reactor at full power. Equation (7.53) indicates that as the volume and therefore $\tilde{H}/M$ increases, the nonleakage probability becomes closer to one; that is, the leakage probability decreases. With a reactor's size and neutron nonleakage probability determined, the fuel enrichment—which has very little effect on the migration length—is adjusted to obtain the desired value of k_∞.

Chapter 8 details the thermal and hydraulic properties of reactor cores that determine the power densities achievable in reactor lattices as well as the coupling between neutronic and thermal-hydraulic design. In the remainder of this chapter we examine first reactor reflectors and their effects on multiplication and flux distribution and then the neutron poisons used to control reactivity and their interactions with the spatial flux distributions.

7.5 Reflected Reactors

Reflectors derive their name from the fact that some fraction of the neutrons that escape the core will make a sufficient number of scattering collisions in the diffusing reflector material to turn them around such that they reenter the core—that is, they are reflected back into to it. A reflector thus reduces the fraction of neutrons leaking from the reactor. Figure 7.4 shows schematic diagrams of cores with axial and radial reflectors. Reflectors have their largest impacts on smaller cores, where the leakage probability is significant. As we shall illustrate, a reflector's importance diminishes as the size of a reactor—measured by $\tilde{H}/M$—becomes larger.

Equation (7.9) for the neutron distribution within a uniform core remains valid for reflected reactors. If both axial and radial reflectors are employed, however, the separation of variables technique applied in Section 7.3 is no longer applicable, and more advanced

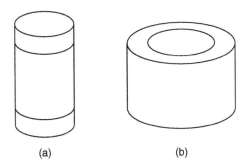

(a)   (b)

FIGURE 7.4   Reflected reactor cores. (a) Axial reflector, (b) radial reflector.

mathematical methods are required. If only an axial or a radial reflector—but not both—is present, the solution again results from inserting Eqs. (7.16) and (7.18) into Eq. (7.11). However the boundary conditions are different, causing the height H of the critical core to decrease in the case of an axial reflector, and the core radius R to decrease in the case of a radial reflector. Conversely, if a reflector is added and the core dimensions are not reduced, the multiplication will increase, and this must be compensated by decreasing k_∞, for example, by reducing the fuel enrichment or adding a neutron absorber to the core. We treat only the axial reflector here; the effects would be analogous for a radial reflector.

Axial Reflector Example

Because no fissionable material is present in a reflector, Eq. (7.3) reduces to

$$\nabla^2 \phi - \frac{1}{\widehat{M}^2}\phi = 0, \tag{7.54}$$

where we have replaced the diffusion length with $\widehat{M}$, the migration length of the reflector material. In cylindrical geometry the ∇^2 operator in Eq. (7.54) takes the same form as in Eq. (7.9). Hence

$$\frac{1}{r}\frac{d}{dr}r\frac{d}{dr}\phi + \frac{d^2}{dz^2}\phi - \frac{1}{\widehat{M}^2}\phi = 0. \tag{7.55}$$

We again separate variables

$$\phi(r,z) = \psi(r)\zeta(z) \tag{7.56}$$

and divide the result by $\psi\zeta$ to yield

$$\frac{1}{\psi r}\frac{d}{dr}r\frac{d}{dr}\psi + \frac{1}{\zeta}\frac{d^2}{dz^2}\zeta - \frac{1}{\widehat{M}^2} = 0. \tag{7.57}$$

The radial flux distribution ψ must satisfy Eq. (7.18), with $C_2' = 0$, and Eq. (7.19) once again determines B_r since ψ must meet the zero flux boundary condition at $r = \tilde{R}$ in the reflector as well as the core. Using Eq. (7.14) to eliminate ψ, and defining

$$\alpha^2 = B_r^2 + \frac{1}{\widehat{M}^2} \tag{7.58}$$

reduces Eq. (7.57) to

$$\frac{d^2}{dz^2}\zeta - \alpha^2\zeta = 0, \tag{7.59}$$

which has a solution of the form

$$\zeta(z) = C_1''\exp(\alpha z) + C_2''\exp(-\alpha z). \tag{7.60}$$

Alternately, using the definitions of sinh and cosh given in Appendix A, we may replace this expression with

$$\zeta(z) = C_1'''\sinh(\alpha z) + C_2'''\cosh(\alpha z). \tag{7.61}$$

We next add a reflector of height a to the top and bottom of the core. Adding the reflector reduces the height of the critical reactor from $\tilde{H}$ to a yet to be determined value of H'; thus the boundary condition at the top of the reflector is $\zeta(H'/2 + a) = 0$. This boundary condition removes one of the arbitrary coefficients from Eq. (7.61). After some algebra, the result may be shown to be

$$\zeta(z) = C'\sinh[\alpha(H'/2 + a - z)]. \tag{7.62}$$

The solution in the core once again takes the form of Eq. (7.16) with $C_1 = 0$, because the solution still must be symmetric about the core midplane. However, the axial buckling, which we now denote as B_z', is yet to be determined since we no longer employ $\chi(\pm\tilde{H}/2) = 0$, the axial boundary condition for a bare reactor. Thus in the core we have

$$\chi(z) = C_2\cos(B_z'z) \tag{7.63}$$

We are now prepared to apply the interface conditions specified by Eqs. (6.42) and (6.43) at the core–reflector interface: continuity of flux,

$$\chi(H'/2) = \zeta(H'/2), \tag{7.64}$$

and of current

$$D\frac{d}{dz}\chi(z)\bigg|_{H'/2} = \hat{D}\frac{d}{dz}\zeta(z)\bigg|_{H'/2}, \tag{7.65}$$

where $\hat{D}$ is the reflector diffusion coefficient. Inserting Eqs. (7.62), and (7.63) into Eqs. (7.64) and (7.65) we obtain, respectively,

$$C_2\cos(B'_z H'/2) = C'\sinh(\alpha a) \tag{7.66}$$

and

$$B'_z D C_2 \sin(B'_z H'/2) = \alpha \hat{D} C' \cosh(\alpha a). \tag{7.67}$$

Taking the ratio of these equations then yields

$$B'_z D \tan(B'_z H'/2) = \alpha \hat{D} \coth(\alpha a). \tag{7.68}$$

For thick reflectors (i.e., ones which are several diffusion lengths thick) the fraction of neutrons that escapes from the outer reflector surface becomes negligible. We may then approximate the reflector as infinite, taking $a \to \infty$. Because $\coth(\infty) = 1$, Eq. (7.68) reduces to

$$B'_z \tan(B'_z H'/2) = \alpha \hat{D}/D, \tag{7.69}$$

or, solving for H',

$$H' = \frac{2}{B'_z} \arctan\left(\alpha \frac{\hat{D}}{B'_z D}\right). \tag{7.70}$$

Note that once the radial dimension of the reactor and the reflector material properties have been specified, α, given by Eq. (7.58), and $\hat{D}$ are specified. For simplicity we assume that D is fixed. Then of the two remaining quantities, B'_z and H', we may fix one and determine the other.

Reflector Savings and Flux Flattening

We proceed by applying the foregoing formulas to two situations. In the first we specify $B'_z \to B_z = \pi/\tilde{H}$. This is equivalent to stating that the core composition is the same as it would be for a bare core of length $\tilde{H}$. Replacing the buckling in Eq. (7.70) then gives

$$H' = \frac{2\tilde{H}}{\pi} \arctan\left(\alpha \frac{\tilde{H}\hat{D}}{\pi D}\right), \tag{7.71}$$

which may be shown to be less than $\tilde{H}$. The reduction in the half core height of the critical reactor is defined as the axial reflector savings: $\delta_z = \frac{1}{2}(\tilde{H} - H')$. For a thick reflector, it is approximately $\delta_z \approx \hat{M}D/\hat{D}$. Figure 7.5 provides a comparison of the flux distributions for a bare and a reflected reactor in which the core composition is held constant and the axial dimension reduced to maintain criticality. The treatment of the radial reflector is analogous, with the complication that Bessel functions are involved. Applying the thick reflector approximation to the radial reflector savings defined by $\delta_r = \tilde{R} - R'$, we would obtain $\delta_r \approx \hat{M}D/\hat{D}$, again for a thick reflector.

The second situation arises from specifying that the core height remain constant by taking $H' = \tilde{H}$. Solving Eq. (7.70) then yields a value $B'_z < \pi/\tilde{H}$. Consequently, replacing B_z with the smaller value B'_z in Eq. (7.13) reduces the overall buckling in Eq. (7.6). Since the denominator decreases with the addition of the reflector, the value of k_∞ must also be decreased—most likely by reducing the fuel enrichment—in order for the multiplication to remain unchanged. Figure 7.6 compares the normalized flux distributions for bare and reflected

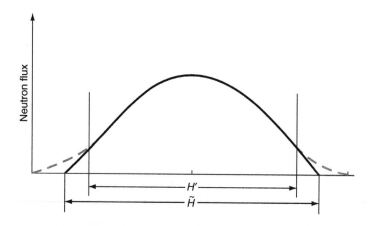

FIGURE 7.5 Axial flux distributions for bare (—) and reflected (- -) reactors with the same core composition.

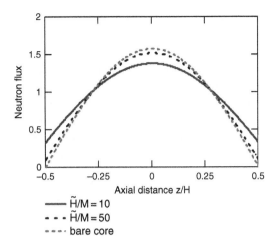

FIGURE 7.6 Comparison of bare and axial reflected reactor flux distributions with the same core height.

reactors in which the height is held constant and the composition changed to maintain criticality.

To recapitulate, adding a reflector to a reactor allows either the volume of the reactor or the value of k_∞ to be reduced, or some combination of the two. Figure 7.6 also illustrates a third effect: Adding a reflector flattens the flux distribution, thus lowering the ratio of peak to average flux. These effects, however, become less significant as the size of the reactor—measured in migration lengths—increases. Clearly, the reflector savings amounts to a smaller fraction of the core dimension as its size becomes larger. Likewise, if the core dimensions are held constant, adding a reflector has a smaller effect on the multiplication of a large reactor than on a small one. Table 7.2 compares the maximum and minimum flux for two reflected reactors with height to migration length ratios of $\tilde{H}/M = 10$ and $\tilde{H}/M = 50$ to a bare reactor, with each of the three average fluxes normalized to one. The numbers clearly indicate the reflector's diminishing effect on flux flattening as the size of the reactor increases.

TABLE 7.2
Flux Characteristics for axial Reflected and Bare Reactors

Reactor	Maximum ϕ	Minimum ϕ	Average ϕ
Reflected, $\tilde{H}/M = 10$	1.373	0.323	1.00
Reflected, $\tilde{H}/M = 50$	1.515	0.091	1.00
Bare	1.571	0.000	1.00

7.6   Control Poisons

Control poisons are neutron absorbers that are deliberately included in a reactor core. They may take the form of control rods, of soluble poisons dissolved in liquid coolants, or of so-called burnable poisons permanently embedded in the fuel or other core constituents. Poisons serve a number of purposes. Control rods are inserted or withdrawn to control the value of k as needed for startup, shutdown, and changes in power level. They may also be used to keep the reactor critical at a constant power by compensating for fuel depletion, fission product buildup, temperature changes, or other phenomena that affect the multiplication. Control poisons affect both the multiplication and the flux distribution of a core; thus we must consider both.

   We begin our analysis of control poisons with Eq. (7.3), assuming for simplicity that we have a uniform unreflected reactor to which we add a control poison. We further assume that the poison has no effect on the fission cross section, and that its small effects on the diffusion coefficient can be ignored. Its primary effect will be an increase in the absorption cross section. We designate this increase by letting $\Sigma_a \to \Sigma_a + \delta\Sigma_a$. Here, the additional absorption, $\delta\Sigma_a$, may be uniform across the core, as in the case of the boron absorber that is added to the coolant of pressurized water reactors, or it may be localized as in the form of one or more control rods. Burnable poisons typically are distributed throughout the core but in an optimized nonuniform manner that is beyond the scope of the following analysis.

Reactivity Worth

Before examining specific control rod configurations we derive a general expression for the reactivity decrease resulting from the addition of a neutron absorber. Such a decrease frequently is referred to as the worth of the absorber. We designate k and ϕ as the multiplication and flux distribution before the neutron poison is inserted and k' and ϕ' as the corresponding values following insertion. Thus the addition of the absorber causes Eq. (7.3) to be replaced by

$$D\nabla^2\phi' + \frac{1}{k'}\nu\Sigma_f\phi' - (\Sigma_a + \delta\Sigma_a)\phi' = 0. \qquad (7.72)$$

We multiply by ϕ and integrate over the reactor volume, V:

$$D\int\phi\nabla^2\phi'dV + \frac{1}{k'}\int\phi\nu\Sigma_f\phi'dV - \int\phi(\Sigma_a + \delta\Sigma_a)\phi'dV = 0. \qquad (7.73)$$

Likewise, we multiply Eq. (7.3) by ϕ' and perform the same volumetric integration to obtain

$$D \int \phi' \nabla^2 \phi \, dV + \frac{1}{k} \int \phi' \nu \Sigma_f \phi \, dV - \int \phi' \Sigma_a \phi \, dV = 0. \qquad (7.74)$$

Subtracting Eq. (7.74) from Eq. (7.73) then yields

$$D \int (\phi \nabla^2 \phi' - \phi' \nabla^2 \phi) dV + \left(\frac{1}{k'} - \frac{1}{k} \right) \int \phi \nu \Sigma_f \phi' \, dV - \int \phi \delta \Sigma_a \phi' \, dV = 0.$$

$$(7.75)$$

Next we demonstrate that the first integral on the left vanishes. Noting that

$$\vec{\nabla} \cdot (\phi \vec{\nabla} \phi') = \phi \nabla^2 \phi' + (\vec{\nabla} \phi) \cdot (\vec{\nabla} \phi') \qquad (7.76)$$

and

$$\vec{\nabla} \cdot (\phi' \vec{\nabla} \phi) = \phi' \nabla^2 \phi + (\vec{\nabla} \phi') \cdot (\vec{\nabla} \phi), \qquad (7.77)$$

we substitute these identities into the first integral of Eq. (7.75) and then use the divergence theorem to convert the volume integral to an integral over the reactor's outer surface area A:

$$\int (\phi \nabla^2 \phi' - \phi' \nabla^2 \phi) dV = \int \vec{\nabla} \cdot (\phi \vec{\nabla} \phi' - \phi' \vec{\nabla} \phi) dV$$

$$= \int \hat{n} \cdot (\phi \vec{\nabla} \phi' - \phi' \vec{\nabla} \phi) dA = 0. \qquad (7.78)$$

The surface integral vanishes since both the flux before and after the change in absorption, that is ϕ and ϕ', must vanish on the extrapolated surface A.

With its first term eliminated, Eq. (7.75) reduces to

$$\left(\frac{1}{k'} - \frac{1}{k} \right) \int \phi \nu \Sigma_f \phi' \, dV = \int \phi \delta \Sigma_a \phi' \, dV. \qquad (7.79)$$

Suppose the reactor is initially critical, then $k = 1$ and the reactivity following control insertion will be $\rho = (k' - 1)/k'$, reducing Eq. (7.79) to

$$\rho = - \int \phi \delta \Sigma_a \phi' \, dV \bigg/ \int \phi \nu \Sigma_f \phi' \, dV. \qquad (7.80)$$

We have made no approximations in obtaining this equation. Moreover, if the added absorption is uniform over the core, the cross sections may be pulled outside the integrals, yielding for the reactivity,

$$\rho = -\delta\Sigma_a/\nu\Sigma_f = -\frac{1}{k_\infty}\delta\Sigma_a/\Sigma_a. \tag{7.81}$$

If the control material is localized within some volume of the reactor, the distribution of the perturbed flux ϕ' must be treated explicitly. Provided the perturbation to the flux is small, however, we may write $\phi' = \phi + \delta\phi$ and ignore the size of $\delta\phi$ relative to ϕ. The result is called the first-order perturbation approximation:

$$\rho = -\int \delta\Sigma_a\phi^2 dV \Big/ \int \nu\Sigma_f\phi^2 dV. \tag{7.82}$$

Partially Inserted Control Rod

Equation (7.82) serves as a basis for estimating the reactivity worth of a partially inserted control rod, provided the reactivity associated with it is not large enough to significantly distort the flux distribution. Consider a uniform cylindrical reactor and begin by rewriting the numerator of Eq. (7.82) explicitly in cylindrical coordinates:

$$\rho = -\int_{-H/2}^{H/2}\int_0^{2\pi}\int_0^R \delta\Sigma_a\phi^2 r\,dr\,d\omega\,dz \Big/ \int\int \nu\Sigma_f\phi^2 dV. \tag{7.83}$$

We evaluate the numerator only over that volume occupied by the control rod. Let $\delta\Sigma_a = \Sigma_{ac}$ in that volume, and $\delta\Sigma_a = 0$ elsewhere. Assume that the rod is located at a distance r from the center of the core and is inserted into the core a distance x from its top as indicated in Fig. 7.7. We continue to assume that the flux is a function only of r and z, and not of the azimuthal angle ω. If the cross-sectional area of the rod, say, a_c, is small we can ignore the r variation of the flux over its diameter and approximate the above expression as

$$\rho_{r,x} = -a_c\Sigma_{ac}\int_{H/2-x}^{H/2}\phi^2(r,z)dz \Big/ \int\int \nu\Sigma_f\phi^2(r,z)dV. \tag{7.84}$$

For the uniform core with the flux described by Eq. (7.21), we obtain

$$\rho_{r,x} = -a_c\Sigma_{ac}\frac{C^2 J_0^2(2.405r/\tilde{R})}{\int \nu\Sigma_f\phi^2(r,z)dV}\int_{\tilde{H}/2-x}^{\tilde{H}/2}\cos^2(\pi z/\tilde{H})dz. \tag{7.85}$$

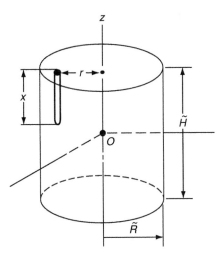

FIGURE 7.7 Control rod inserted to a depth x at a distance r from a core centerline.

This expression simplifies when normalized to a fully inserted control rod $(x = \tilde{H})$ at the same radial location. Dividing $\rho_{r,x}$ by $\rho_{r,H}$ we obtain

$$\rho_{r,x} = \frac{\int_{\tilde{H}/2-x}^{\tilde{H}/2} \cos^2(\pi z/\tilde{H}) dz}{\int_{-\tilde{H}/2}^{\tilde{H}/2} \cos^2(\pi z/\tilde{H}) dz} \rho_{r,H} = \left[\frac{x}{\tilde{H}} - \frac{1}{2\pi} \sin(2\pi x/\tilde{H}) \right] \rho_{r,H}. \quad (7.86)$$

Figure 7.8 shows the control rod worth, that is, the negative reactivity created by the rod, normalized to the fully inserted rod. Note that the worth changes most rapidly when the tip is near the midplane of the core, where the flux is largest. Control rod worth also

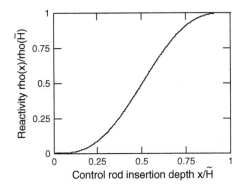

FIGURE 7.8 Normalized control rod worth vs insertion depth.

varies with the radial position of the core. For example, if we compare a rod situated at a radius r to one located along the core's centerline, because $J_0(0) = 1$, we obtain

$$\rho_{r,x} = J_0^2(2.405r/\tilde{R})\rho_{0,x}. \tag{7.87}$$

Control Rod Bank Insertion

Normally, control rods are grouped into banks, which may be inserted or withdrawn in unison. As such, rod banks have large enough effects on reactivity and on the flux distribution that the perturbation technique described above is no longer applicable. To model the reactivity worth of a bank of control rods, we again consider a uniform cylindrical reactor, and specify a height to diameter ratio of one. To simplify the derivation we take the origin of our r–z coordinates at the base of the reactor, as indicated in Fig. 7.9, and assume that the bank is inserted a distance x from the top of the reactor as indicated. Our starting point is therefore Eq. (7.4), but k_∞ and M are no longer constants; they are functions of z, taking different values in the rodded and unrodded volumes of the core. Since the core is uniform in the radial direction, however, we may again employ separation of variables as in Eq. (7.11). Inserting these separated variables into Eq. (7.5) yields

$$\frac{1}{\psi r}\frac{d}{dr}r\frac{d}{dr}\psi + \frac{1}{\chi}\frac{d^2}{dz^2}\chi + \frac{k_\infty/k - 1}{M^2} = 0, \tag{7.88}$$

where we have replaced the diffusion with the migration length. The first term on the left must be a constant, since the remaining two

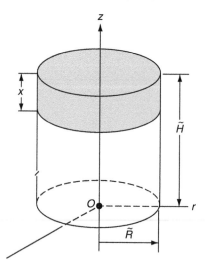

FIGURE 7.9   Cylindrical reactor with a control rod bank inserted to depth x.

terms vary only in z. We set it equal to $-B_r^2$ given by Eq. (7.19), thus satisfying Eq. (7.14) and meeting the boundary condition that $\phi(\tilde{R}, z) = 0$. Thus Eq. (7.88) reduces to

$$\frac{d^2}{dz^2}\chi + \left(\frac{k_\infty/k - 1}{M^2} - B_r^2\right)\chi = 0. \tag{7.89}$$

To account for the absorption in the control rod bank we subtract δk_∞ from the infinite medium multiplication in the axial region of the core occupied by the rod bank. Hence for that region,

$$k_\infty \to k_\infty - \delta k_\infty = k_\infty(1 - \rho_b), \tag{7.90}$$

where $\rho_b = \delta k_\infty/k_\infty$ is the reactivity worth of the bank when inserted the entire length of the core. For simplicity we assume that the presence of the control rods has no appreciable effect on the migration length. Equation (7.89) takes two forms for unrodded and rodded volumes (u and r, respectively). With the rod bank inserted to a depth x from the top of the core:

$$\frac{d^2}{dz^2}\chi_u + \alpha^2\chi_u = 0, \quad 0 \le z \le \tilde{H} - x, \tag{7.91}$$

and

$$\frac{d^2}{dz^2}\chi_r + (\alpha^2 - \beta^2)\chi_r = 0, \quad \tilde{H} - x \le z \le \tilde{H}, \tag{7.92}$$

where

$$\alpha^2 = \frac{1}{M^2}\left(\frac{k_\infty}{k} - 1\right) - B_r^2 \tag{7.93}$$

and

$$\beta^2 = \frac{1}{M^2}\frac{k_\infty}{k}\rho_b. \tag{7.94}$$

The solutions must meet the boundary conditions $\chi_u(0) = 0$ and $\chi_r(\tilde{H}) = 0$ as well as the interface conditions

$$\chi_u(\tilde{H} - x) = \chi_r(\tilde{H} - x) \tag{7.95}$$

and

$$\frac{d}{dz}\chi_u(z)\Big|_{\tilde{H}-x} = \frac{d}{dz}\chi_r(z)\Big|_{\tilde{H}-x}, \qquad (7.96)$$

where we have made the assumption that the diffusion coefficients in the rodded and unrodded core regions are the same. Solutions meeting the boundary conditions at $z = 0$ and $z = \tilde{H}$ may be shown to be

$$\chi_u(z) = C\sin(\alpha z) \qquad (7.97)$$

and

$$\chi_r(z) = \begin{cases} C'\sin\left[\sqrt{\alpha^2 - \beta^2}(\tilde{H} - z)\right], & \alpha^2 > \beta^2 \\ C'\sinh\left[\sqrt{\beta^2 - \alpha^2}(\tilde{H} - z)\right], & \alpha^2 < \beta^2. \end{cases} \qquad (7.98)$$

Application of the interface conditions, Eqs. (7.95) and (7.96), leads to two additional equations relating C and C'. Taking their ratio yields the transcendental equation

$$\alpha \cot[\alpha(\tilde{H} - x)] = -\sqrt{\alpha^2 - \beta^2}\cot\left(\sqrt{\alpha^2 - \beta^2}x\right), \quad \alpha^2 > \beta^2, \qquad (7.99)$$

or

$$\alpha \cot[\alpha(\tilde{H} - x)] = -\sqrt{\beta^2 - \alpha^2}\coth\left(\sqrt{\beta^2 - \alpha^2}x\right), \quad \alpha^2 < \beta^2. \qquad (7.100)$$

Given the reactor's material properties, along with the definitions of α and β, these transcendental equations can be solved numerically for k.

The two limiting values of k reduce to be what is expected. For the rods out case, Eq. (7.97) is applicable over the entire core. Thus we must have $\chi_u(\tilde{H}) = 0$, from which it follows that $\alpha_u^2 = (\pi/\tilde{H})^2 = B_z^2$, so that Eq. (7.93) yields

$$k_u = \frac{k_\infty}{1 + M^2(B_r^2 + B_z^2)}. \qquad (7.101)$$

Similarly, with the control rods fully inserted, Eq. (7.98) holds over the entire core, and we must have $\chi_r(0) = 0$. Only the first of the pair can meet this condition, and we obtain $\alpha_r^2 - \beta_r^2 = (\pi/\tilde{H})^2 = B_z^2$, from which Eqs. (7.93) and (7.94) yield $k_r = (1 - \rho_b)k_u$. Next we consider situations where the rod bank is partially inserted to a distance x.

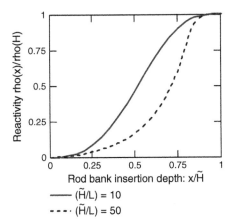

FIGURE 7.10 Normalized control rod bank worth vs insertion depth for core of heights $\tilde{H}/M = 10$ and $\tilde{H}/M = 50\,H$.

Figure 7.10 is a plot of the reactivity versus insertion length for two cores, both with height to diameter ratios of one, and both with rod bank worths of $\rho_b = 0.02$. The ratio of core dimension to migration length, $\tilde{H}/M$, is central to understanding the reactivity curves and flux distributions. A core with a smaller value of $\tilde{H}/M$ is referred to as neutronically tightly coupled, because disturbances, such as control rod bank insertions, travel across it in relatively few migration lengths, and therefore over relatively few neutron generations. In a loosely coupled core—one with a large $\tilde{H}/M$ ratio—the converse is true: Many neutron generations are required to propagate a disturbance across the reactor. The net effect is that in the more tightly coupled core distribution of the neutron flux deviates less from that of a uniform core, and the pattern of control rod worth is closer to that of Eq. (7.86); this can be observed by comparing the $\tilde{H}/M = 10$ curve to Fig. 7.8, which assumes no disturbance to the flux distribution. In contrast, the $\tilde{H}/M = 50$ curve is quite skewed; the rod bank has little effect until it is inserted more than halfway into the core.

The effects on flux distribution are also pronounced. As Fig. 7.11a indicates, as the rod bank is inserted into the more tightly coupled core, with $\tilde{H}/M = 10$, the perturbed flux does not deviate greatly from the standard axial cosine distribution. For the more loosely coupled core, however, the flux is pushed toward the bottom of the core as the rods are inserted. This is illustrated for the $\tilde{H}/M = 50$ case shown in Fig. 7.11b. Note that all the curves in Figs. 7.11a and 7.11b are normalized to the same average flux, thus illustrating that the flux peaking can become quite severe in the more loosely coupled core.

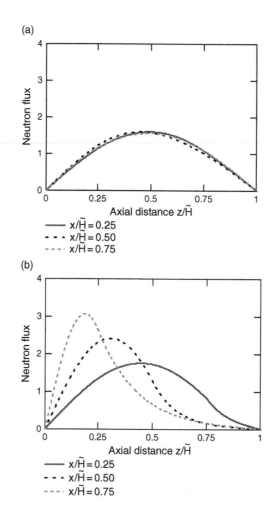

FIGURE 7.11   Normalized axial flux distributions for control rod bank insertion depths of $x/\tilde{H} = 0.25$, 0.50, and 0.75. (a) $\tilde{H}/\tilde{M} = 10$, (b) $\tilde{H}/M = 50$.

A reactor's power density (i.e., power per unit volume) is limited by thermal-hydraulic considerations. Thus for a given class of reactors, higher powers imply larger volumes, and therefore larger values of $\tilde{H}/M$, and more loosely coupled cores.

Comparing Tables 4.1 and 7.1 we see that both achievable average power densities and migration lengths vary greatly between reactor designs. Generally, thermal reactors are more loosely coupled than fast reactors, and they may have values of $\tilde{H}/M$ substantially exceeding 50. Thus prevention of excessive power peaking from control rod movements, refueling patterning, and other phenomena becomes an increasing concern. In the following chapter we take up the coupling between neutronics and the removal of heat from power reactors that bears heavily on these issues.

Bibliography

Bell, George I., and Samuel Glasstone, *Nuclear Reactor Theory*, Van Nostrand-Reinhold, NY, 1970.

Duderstadt, James J., and Louis J. Hamilton, *Nuclear Reactor Analysis*, Wiley, NY, 1976.

Henry, Allen F., *Nuclear-Reactor Analysis*, MIT Press, Cambridge, MA, 1975.

Lamarsh, John R., *Introduction to Nuclear Reactor Theory*, Addison-Wesley, Reading, MA, 1972.

Meghreblian, R. V., and D. K. Holmes, *Reactor Analysis*, McGraw-Hill, NY, 1960.

Ott, Karl O., and Winfred A. Bezella, *Introductory Nuclear Reactor Statics*, 2nd ed., American Nuclear Society, 1989.

Stacey, Weston M., *Nuclear Reactor Physics*, Wiley, NY, 2001.

Problems

7.1. The material composition for the core of a large reactor yields $k_\infty = 1.02$ and $M = 25$ cm.

   a. Calculate the critical volume for a bare cylinder with a height to diameter ratio of one.
   b. Calculate the critical volume of a bare sphere.

   Which of the two volumes did you expect to be larger? Why?

7.2. Determine the height to diameter ratio of a bare cylindrical reactor that will lead to the smallest critical mass.

7.3. Critical assemblies for studying the properties of fast reactors are sometimes built in halves as shown in the figure. The two halves are maintained in subcritical states by separating them with a sufficient distance that neutronic coupling between the two is negligible; they are then brought together to form a critical assembly. Suppose the core composition under investigation has an infinite medium multiplication of 1.36 and a migration length of 18.0 cm. The assembly is configured with a height to diameter ratio of one $(H = D)$. Neglecting extrapolation distances,

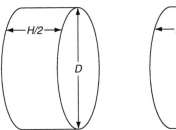
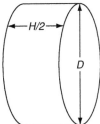

   a. Determine the dimensions required to make the assembly exactly critical when the two halves are brought into contact.
   b. Determine the value of k for each of the halves when they are isolated from each other.

7.4. A sodium-cooled fast reactor is to be built with the composition given in problem 4.3 with a height to diameter ratio of 0.8. If the reactor, which is bare, is to have a value of $k = 1.005$ when no control rods are inserted, what should the value of the reactor's diameter be?

7.5. A cylindrical tank is constructed for storage of liquids containing fissionable material. The tank has a diameter of 0.90 m, and it is surrounded by a nonreflecting neutron absorber. Material with $k_\infty = 1.16$ and $M = 7.0$ cm is poured into the tank. Neglecting extrapolation lengths,

   a. To what height can the tank be filled before it becomes critical?
   b. Estimate the maximum value of k_∞ permissible (M remaining unchanged) if it must be guaranteed that criticality will not be reached no matter to what height the tank is filled.
   c. It is decided to reduce the diameter of the tank so that the material in part a can never reach criticality. Estimate the reduced diameter.

7.6. Consider a critical reactor that is a cube with extrapolated side length a.

   a. With the origin at the center, apply separation of variables in three-dimensional Cartesian geometry to show that the flux distribution is

$$\phi(x, y, z) = C \cos(\pi x/a) \cos(\pi y/a) \cos(\pi z/a).$$

   b. Find C in terms of the reactor power, volume, and $\gamma \Sigma_f$.
   c. Determine the reactor's buckling.
   d. Suppose that $a = 2.0$ m and $M = 20$ cm. Determine the value of k_∞ required to obtain criticality (i.e., $k = 1.0$).

7.7.* A cylindrical fast reactor has a volume of 1.4 m^3 and a migration length of 20 cm. For height to diameter ratios between 0.5 and 2.0 make plots on the same graph of the following:

   a. The nonleakage probability P_{NL}.
   b. The value of k_∞ required for the reactor to be critical.

7.8. A critical bare cylindrical reactor has a height to diameter ratio of one and a migration length of 7.5 cm. The core volume is 15 m^3. To simplify analysis, an engineer replaced the cylinder with a sphere of the same volume.

a. What is the sign and magnitude of the error in the multiplication caused by this simplification?

b. If the reactor has a larger volume of $30\,\text{m}^3$, will the error be larger or smaller than in part a? Justify your result.

7.9. Show that $C_1 = 0$ in Eq. (7.16) from the boundary conditions $\chi(\pm\tilde{H}/2) = 0$ without employing the symmetry condition $\chi(z) = \chi(-z)$.

7.10. Express C_1''' and C_2''' in Eqs. (7.61) in terms of C_1'' and C_2'' in Eq. (7.60).

7.11. Apply the boundary condition $\zeta(\tilde{H}'/2 + a) = 0$ to determine C_2''' in terms of C_1''' in Eq. (7.61). Then determine C' in Eq. (7.62) in terms of C_1''' and C_2'''.

7.12. Consider the situation when the spherical system discussed in Section 6.7 is critical. Determine the ratio of maximum to average flux in the sphere.

7.13. A spherical reactor of radius R is surrounded by a reflector that extends to $r = \infty$. L and D are the same for core and reflector. Find the criticality equation relating k_∞, R, L, and D.

7.14. A spherical reactor is constructed with an internal reflector with parameters and D and Σ_a^r and extending $0 \le r \le R$. The annular core, with parameters D, Σ_a, and k_∞ (>1), extends $R \le r \le 2R$.

a. Find the criticality condition (neglecting the extrapolation distance).

b. Sketch the flux distribution for $0 \le r \le 2R$.

7.15. Show from Eq. (7.71) that the reflector savings is approximated by $\delta_z \approx \widehat{M}D/D$ for a thick reflector.

7.16. An infinite slab reactor (extending to infinity in the y and z directions) has a thickness of $2a$ with vacuum on either side. The properties for material 1 occupying $0 \le x \le a$ are $k_\infty^1 = k_\infty$, $D_1 = D$, and $\Sigma_a^1 = \Sigma_a$ and those for material 2 occupying $a \le x \le 2a$ are $k_\infty^2 = 0$, $D_2 = D$, and $\Sigma_a^2 = 0$. Neglecting extrapolation distances,

a. Find a criticality equation relating a, k_∞, D, and Σ_a.

b. Sketch the flux between 0 and $2a$.

7.17. Apply the interface conditions at the tip of the control rod bank to show that Eqs. (7.97) and (7.98) yield the criticality condition given by Eqs. (7.99) and (7.100).

7.18. Beginning with Eq. (7.98) prove that with the control rod bank fully inserted.

a. $\alpha_r^2 - \beta_r^2 = B_z^2$

b. $k_r = (1 - \rho_b)k_u$

CHAPTER 8

Energy Transport

8.1 Introduction

The preceding three chapters concentrated on the time and space distributions of neutrons in a reactor core, leading to Eq. (7.30), which shows that in a critical reactor the flux level is proportional to the reactor power. At very low power any level can be chosen, and the equation still holds. However, for higher powers, at the levels typically found in power reactors, two important limitations come into play. First, the energy must be transported out of the core without overheating the fuel, coolant, or other constituents; these thermal limits determine the maximum power at which a reactor can operate. Second, as temperatures rise, the densities of the core materials change at different rates, and other temperature-related phenomena occur. These affect the multiplication, causing temperature-related reactivity feedback effects to ensue.

This chapter examines the energy transport from reactor cores, defining power density and related quantities that determine temperature distributions. These quantities allow the thermal limits imposed on reactor operation to be examined in the later part of the chapter. Chapter 9 combines the temperature distributions with the neutron physics of reactor lattices to examine reactivity feedback effects.

8.2 Core Power Distribution

The interaction of neutron physics with heat transport requirements may be understood in broad terms as follows. Let P signify the reactor power and V the core volume. Then

$$\bar{P}''' = P/V \tag{8.1}$$

199

defines the core-averaged power density. The ratio of maximum to average power density is the power peaking factor:

$$F_q = P'''_{\max}/\bar{P}'''. \tag{8.2}$$

Eliminating $\bar{P}'''$ between these definitions offers some insight into the interdisciplinary nature of reactor core design:

$$P = \frac{P'''_{\max}}{F_q} V. \tag{8.3}$$

Reactors are normally designed to produce a specified amount of power, while with other variables held constant the cost of construction rises dramatically with the core volume. Thus maximizing the ratio $P'''_{\max}/F_q$ is a central optimization problem of core design. The achievable maximum power density is dependent primarily on materials properties and the temperatures and pressures that can be tolerated by fuel, coolant, and other core constituents. Minimizing the peaking factor falls much more into the domain of reactor physics, for nonuniform distributions of fuel enrichment, the positioning of control rods and other neutron poisons, as well as other neutronic considerations largely determine the value of F_q. The core volume that is ultimately selected also has reactor physics repercussions, most importantly on the core-averaged fuel enrichment and the nonleakage probability. Table 8.1 displays representative properties for the major classes of power reactors.

Finite Cylindrical Core

The distribution of power density is central to the interaction of reactor physics with thermal-hydraulic phenomena. The power density in watts/cm^3, kW/liter or MW/m^3 at any point $\vec{r}$ in the reactor is given by

$$P'''(\vec{r}) = \gamma \Sigma_f(\vec{r}) \phi(\vec{r}), \tag{8.4}$$

where γ is the number of W-s/fission and $\Sigma_f \phi$ is the number of fissions/cm^3/s. The fission cross section and the flux are spatial averages over lattice cells of fuel, coolant, and other core constituents, with the averages determined by the methods discussed in Chapter 4. We consider specifically a cylindrical reactor of height H and radius R, and take the origin at $r=z=0$, the center of the reactor. We further assume that the power distribution, and therefore Σ_f and ϕ in Eq. (8.4), are separable functions of r and z. With this restriction we may represent the power density distribution as

$$P'''(r,z) = \bar{P}''' f_r(r) f_z(z). \tag{8.5}$$

TABLE 8.1
3000 MW(t) Power Reactor Approximate Core Properties

	PWR Pressurized-H_2O Reactor	BWR Boiling-H_2O Reactor	PHWR CANDU-D_2O Reactor	HTGR C-Moderated Reactor	SFR Na-Cooled Fast Reactor	GCFR He-cooled Fast Reactor
$\bar{P}'''$ (MW/m^3) average power density	102	56	7.7	6.6	217	115
$\bar{q}'$ (kW/m) average linear heat rate	17.5	20.7	24.7	3.7	22.9	17
V (m^3) core volume	29.4	53.7	390	455	13.8	26.1
H/M height and diameter in migration lengths	43.9	55.0	40.8	68.8	13.5	12.6
N number of fuel pins	51,244	35,474	15,344	97,303	50,365	54,903
P_{NL} nonleakage probability	0.956	0.972	0.950	0.982	0.676	0.644

Source: Data courtesy of W. S. Yang, Argonne National Laboratory.

Since the volume integral

$$\bar{P}''' = \frac{1}{V} \int P'''(\vec{r})dV \tag{8.6}$$

defines the core-averaged power density, inserting $P'''(r,z)$ into this expression determines normalization conditions on $f_r(r)$ and $f_z(z)$. For cylindrical geometry we write the volume integration as

$$\frac{dV}{V} = \frac{2\pi r \, dr \, dz}{\pi R^2 \quad H} \tag{8.7}$$

and obtain

$$\bar{P}''' = \bar{P}''' \frac{2}{R^2} \int_0^R f_r(r) r \, dr \frac{1}{H} \int_{-H/2}^{H/2} f_z(z) dz. \tag{8.8}$$

Equations (8.5) and (8.6) are thus satisfied by the normalization conditions

$$\frac{2}{R^2} \int_0^R f_r(r) r dr = 1 \tag{8.9}$$

and

$$\frac{1}{H} \int_{-H/2}^{H/2} f_z(z) dz = 1. \tag{8.10}$$

The power peaking factor is then the product of radial and axial components, $F_q = F_r F_z$, where the radial and axial peaking factors are

$$F_r = f_r(r)_{\max} \tag{8.11}$$

and

$$F_z = f_z(z)_{\max}. \tag{8.12}$$

Common practice is also to include a local peaking factor F_l to account for fuel element manufacturing tolerances, local control and instrumentation perturbations, and other localized effects on the power density, in which case

$$F_q = F_r F_z F_l. \tag{8.13}$$

In the quest to flatten the power distribution and reduce the peaking factor, two or more radial zones containing fuels of different enrichments frequently serve to decrease the radial peaking factor.

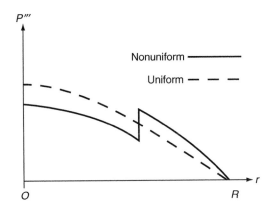

FIGURE 8.1   Radial power distributions for uniform and nonuniform fuel loadings.

The radial power profile might then look as shown in Fig. 8.1, where the discontinuities on the power distribution are due to the discontinuities in the fission cross section. In the axial direction, distortion of the power occurs if partially inserted control rod banks enter from one end of the core as shown in Fig. 7.11, where a control rod bank enters from the top results in a power tilt, causing the flux to be depressed toward the upper end of the core and peaked in the lower half. The result is an increase in F_z. As comparison of Figs. 7.11a and 7.11b indicates, the magnitudes of such distortions tend to grow with the reactors' dimensions, measured in migration lengths.

Uniform Cylindrical Core Example

For a uniform core, the fission cross section in Eq. (8.4) is a constant. Thus the power density is proportional to the flux. Equation (7.30) indicates that for a uniform core the spatial flux dependence is $J_0(2.405r/R)\cos(\pi z/H)$. Thus we take $f_r(r)$ and $f_z(z)$ to have the forms

$$f_r(r) = C_r J_0(2.405r/R) \tag{8.14}$$

and

$$f_z(z) = C_z \cos(\pi z/H), \tag{8.15}$$

where C_r and C_z are normalization coefficients which we determine by inserting these expressions into Eqs. (8.9) and (8.10):

$$C_r \frac{2}{R^2} \int_0^R J_0(2.405r/R)r\,dr = 1 \tag{8.16}$$

and

$$C_z \frac{1}{H} \int_{-H/2}^{H/2} \cos(\pi z/H)dz = 1. \tag{8.17}$$

We have already evaluated integrals identical to these in Eqs. (7.26) and (7.27), yielding

$$f_r(r) = 2.32 J_0(2.405r/R) \tag{8.18}$$

and

$$f_z(z) = 1.57 \cos(\pi z/H). \tag{8.19}$$

Because both the Bessel function and cosine have maximum values of one, the radial and axial peaking factors are

$$F_r = 2.32 \tag{8.20}$$

and

$$F_z = 1.57, \tag{8.21}$$

and Eq. (8.13) yields

$$F_q = 3.63 F_l. \tag{8.22}$$

8.3 Heat Transport

The same lattice structure that profoundly affects the reactor multiplication, as detailed in Chapter 4, largely determines the transport of heat out of the reactor. In what follows we present a simple fuel–coolant model that is applicable to fast reactors and to thermal reactors in which the liquid coolant also serves as the moderator. In thermal reactors using solid moderators, most frequently graphite, a three-region model would be required, taking into account the temperatures in the moderator as well as the coolant. We consider steady state heat transfer in this section before making a brief examination of thermal transients in Section 8.4.

Heat Source Characterization

Consider the cylindrical reactor discussed above, with a height H and a radius R. Taking the origin of the coordinate system at the reactor's center, we designate the properties at a distance z along the axis of a lattice cell whose centerline is located a radial distance r from the core centerline by the arguments (r, z). For simplicity, we assume that the reactor's power distribution is separable in the r and z directions allowing us to describe it with Eq. (8.5).

Let $q'(r, z)$ be the thermal power produced per unit length of a fuel element located in a lattice cell at (r, z); q' is referred to as the *linear heat rate*; it has units of W/cm or kW/m. A related quantity is the *surface heat flux* $q''(r, z)$, measured in W/cm^2 or kW/m^2. It measures the rate of heat flow across the fuel element surface and into the coolant, for a lattice cell located at (r, z). For a cylindrical fuel element with radius a, surface heat flux and linear heat rate are related by

$$q''(r, z) = \frac{1}{2\pi a} q'(r, z). \tag{8.23}$$

Let A_{cell} be the cross-sectional area of a lattice cell containing one fuel element. The thermal power produced per unit volume of core at that point (i.e., the power density) is then

$$P'''(r, z) = q'(r, z)/A_{cell}. \tag{8.24}$$

Combining this expression with Eq. (8.5) allows the linear heat rate to be expressed in terms of the radial and axial power components of the power density:

$$q'(r, z) = A_{cell}\bar{P}'''f_r(r)f_z(z), \tag{8.25}$$

or correspondingly in terms of the reactor power,

$$q'(r, z) = \frac{A_{cell}}{V}Pf_r(r)f_z(z), \tag{8.26}$$

since $\bar{P}''' = P/V$. If the reactor consists of N identical lattice cells, each with a cross-sectional area of A_{cell}, then we may express reactor's volume approximately as

$$V = \pi R^2 H = NA_{cell}H \tag{8.27}$$

and combine this expression with Eq. (8.26) to yield

$$q'(r, z) = \frac{1}{NH}Pf_r(r)f_z(z). \tag{8.28}$$

Note that NH is just the total length of the N fuel elements contained in the reactor core.

Steady State Temperatures

In what follows we develop some approximate expressions for estimating the fuel and coolant temperature distributions in terms of the reactor power. The temperature drop from fuel to coolant is proportional to the linear heat rate:

$$T_{fe}(r,z) - T_c(r,z) = R'_{fe}q'(r,z), \tag{8.29}$$

where $T_{fe}(r,z)$ is the fuel element temperature averaged over the fuel element's cross section, πa^2, and $T_c(r,z)$ is the coolant temperature averaged over the cross-sectional area of the coolant channel associated with the lattice cell. We designate the proportionality constant, R'_{fe}, as the fuel element thermal resistance.

Appendix D contains the derivation of a simple expression for R'_{fe} and shows that for cylindrical fuel elements R'_{fe} is nearly independent of the pin diameter but inversely proportional to the fuel thermal conductivity. The combination of Eqs. (8.28) and (8.29) expresses the temperature drop in terms of the reactor power:

$$T_{fe}(r,z) - T_c(r,z) = R_f P f_r(r) f_z(z), \tag{8.30}$$

where

$$R_f = \frac{1}{NH}R'_{fe} \tag{8.31}$$

defines the reactor core thermal resistance. We may volume-average the fuel and coolant temperatures over the core by applying the integration, defined by Eq. (8.7), to Eq. (8.30):

$$\overline{T}_f - \overline{T}_c = R_f P, \tag{8.32}$$

where we have used the normalization conditions of Eqs. (8.9) and (8.10) to simplify the result.

We proceed by modeling the average and outlet coolant temperatures, $\overline{T}_c$ and $T_o(r)$, under the stipulation that T_i, the inlet temperature, is uniform over the cross-sectional area of the core. The coolant heat balance for a lattice cell located a distance r from the reactor centerline states that the heat added to the coolant is equal to that produced in the fuel element:

$$W_{ch}c_p[T_o(r) - T_i] = \int_{-H/2}^{H/2} q'(r,z')dz', \tag{8.33}$$

where W_{ch} is the mass flow rate in kg/s and c_p is the coolant specific heat at constant pressure, in J/kg K. Inserting Eq. (8.28) expresses the equation's right-hand side in terms of the reactor power. Solving for $T_o(r)$ yields

$$T_o(r) = \frac{1}{Wc_p}\frac{P}{NH}f_r(r)\int_{-H/2}^{H/2} f_z(z)dz + T_i. \qquad (8.34)$$

Next, we define

$$W = NW_{ch} \qquad (8.35)$$

as the core mass flow rate through the N identical channels. Then, using Eq. (8.10) to eliminate $f_z(z)$, we have

$$T_0(r) = \frac{1}{Wc_p}Pf_r(r) + T_i. \qquad (8.36)$$

The average core outlet temperature results from integrating this relationship over the cross-sectional area of the core, that is, by taking $2rdr/R^2$, and applying the normalization of Eq. (8.9):

$$\bar{T}_0 = \frac{1}{Wc_p}P + T_i. \qquad (8.37)$$

The coolant temperature averaged over the core volume depends on details of the axial power distribution. However, if the power distribution does not depart significantly from the axially symmetric condition

$$f_z(z) \approx f_z(-z), \qquad (8.38)$$

a reasonable approximation for the average coolant temperature is

$$\bar{T}_c = \frac{1}{2}(\bar{T}_o + T_i). \qquad (8.39)$$

Using this approximation, we may employ Eq. (8.37) to obtain the average coolant temperature in terms of the known inlet temperature and the reactor power:

$$\bar{T}_c = \frac{1}{2Wc_p}P + T_i. \qquad (8.40)$$

Likewise, inserting this expression into Eq. (8.32) yields an average fuel temperature of

$$\bar{T}_f = \left(R_f + \frac{1}{2Wc_p}\right)P + T_i. \qquad (8.41)$$

Average coolant and fuel temperatures are required for modeling the reactivity feedback effects discussed in the following chapter. Thermal limits, however, are more closely related to maximum fuel and coolant temperatures as well as to the maximum values of the linear heat rate and surface heat flux. The maximum coolant temperature occurs at the core outlet, and may be obtained directly from Eq. (8.36):

$$T_0|_{max} = \frac{1}{Wc_p} PF_r + T_i, \tag{8.42}$$

where F_r is the radial peaking factor defined by Eq. (8.11). The maximum temperature drop from fuel to coolant is determined from Eq. (8.30) by employing the definitions of F_r and F_z:

$$\left[T_{fe}(r,z) - T_c(r,z)\right]\big|_{max} = R_f PF_r F_z. \tag{8.43}$$

Determining $T_{fe}|_{max}$, however, requires that we know the coolant temperature at the point where it occurs. Taking the maximum outlet temperature from Eq. (8.42) would yield too large a value. A closer estimate results from taking the average coolant temperature from the channel with the maximum power output, that is, $F_r \bar{T}_c$. Combining Eqs. (8.41) and (8.43) with this approximation then yields a maximum fuel temperature of

$$T_f|_{max} = \left(R_f F_r F_z + \frac{1}{2Wc_p} F_r\right)P + T_i. \tag{8.44}$$

For liquid-cooled reactors, particularly those utilizing oxide or carbide fuels, the temperature drop from fuel to coolant is much larger than the temperature increase in the coolant. Thus from Eqs. (8.32) and (8.40),

$$\frac{\bar{T}_f - \bar{T}_c}{\bar{T}_c - T_i} = 2Wc_p R_f \gg 1. \tag{8.45}$$

Because $R_f \gg 1/Wc_p$, approximations made in determining the coolant temperature for use in Eq. (8.43) have relatively little impact on maximum fuel temperature. Much more important is the following: The core thermal resistance R_f is derived for T_{fe} averaged over the cross-sectional area of the fuel rod. The temperature at the rod's hottest point, which is along its centerline, determines the limitation on the linear heat rate. As indicated in Appendix D, the centerline temperature may be obtained by replacing R_f in Eq. (8.44) with R_{cl}, the thermal resistance from the centerline to the coolant, which for a cylindrical fuel element is approximately

$$R_{cl} \approx 2 \cdot R_f. \tag{8.46}$$

Pressurized Water Reactor Example

The interplay between neutronics and thermal hydraulics may be demonstrated though the following simplified design analysis of a reactor that is cooled and moderated by pressurized water. The initial design study assumes a uniform cylindrical core with a height to diameter ratio of one, and no significant reflector savings. Assume the following specifications:

Power	$P = 3000\,\text{MW(t)}$
Moderator/fuel ratio	$V_{H_2O}/V_{fuel} = 1.9$
Linear heat rate	$q'\mid_{max} = 400\ \text{W/cm}$
Surface heat flux	$q''\mid_{max} = 125\ \text{W/cm}^2$
Inlet coolant temperature	$T_i = 290°\text{C}$
Outlet coolant temperature	$T_o\mid_{max} = 330°\text{C}$

The reactor's owner determines the power at which the reactor must operate. The moderator to fuel ratio depends primarily on reactor physics considerations. Fuel properties such as thermal conductivity and melting temperature determine the maximum linear heat rate. Surface heat flux in a pressurized water reactor is limited to prevent the possibility of a boiling crisis in which the fuel is insulated from the coolant as the result of the formation of a vapor blanket at the fuel–coolant interface. The coolant inlet temperature results from thermodynamic analysis of the power plant, whereas the outlet temperature must be limited to prevent coolant boiling while operating at the reactor's design pressure.

The foregoing specifications—of which only the moderator to fuel volume ratio depends primarily on neutronics—determine a great many of the reactor's physical characteristics:

a. Fuel radius
b. Lattice pitch
c. Core volume and dimensions
d. Core-averaged power density
e. Number of fuel elements
f. Coolant mass flow rate
g. Mean coolant velocity.

In what follows, we determine each of these in sequence.

The fuel radius is determined from Eq. (8.23):

$$\text{(a)} \quad a = \frac{q'\mid_{max}}{2\pi q''\mid_{max}} = \frac{400}{2\pi \cdot 125} = 0.509 \text{ cm}.$$

Assuming square cells, the lattice pitch is determined from $V_{H_2O}/V_{fuel} = (p^2 - \pi a^2)/\pi a^2$ or

(b) $p = \sqrt{\pi(V_{H_2O}/V_{fuel} + 1)}\, a = \sqrt{2.9\pi}\, 0.509 = 1.536\,\text{cm}.$

For the core volume we evaluate Eq. (8.26) at the point of the maximum linear heat rate to obtain $V = A_{cell}PF_rF_z/q'|_{max}$, where $A_{cell} = p^2$. For a bare, uniform core, Eqs. (8.20) and (8.21) stipulate that $F_r = 2.32$ and $F_z = 1.57$. Thus

$$V = p^2PF_rF_z/q'|_{max} = 1.536^2 \cdot 3000 \cdot 10^6 \cdot 2.32 \cdot 1.57/400$$
$$= 6.445 \cdot 10^7\,\text{cm}^3 = 64.45\,\text{m}^3.$$

For a height to diameter ratio of one, $V = \pi(H/2)^2 H$. Hence

(c) $H = (4V/\pi)^{1/3} = (4 \cdot 6.445 \cdot 10^7/\pi)^{1/3} = 434\,\text{cm} = 4.34\,\text{m}.$

The core-averaged power density is just

(d) $\bar{P}''' = P/V = 3000 \cdot 10^6/6.445 \cdot 10^7 = 46.5\,\text{W/cm}^3 = 46.5\,\text{MW/m}^3.$

The number of fuel elements is determined by dividing the core cross-sectional area by the area of a lattice cell:

(e) $N = \dfrac{\pi R^2}{A_{cell}} = \dfrac{\pi(H/2)^2}{p^2} = \dfrac{\pi(434/2)^2}{1.536^2} = 62{,}702.$

Because the maximum outlet temperature is limited to 330 °C we determine the mass flow rate from Eq. (8.42); taking $c_p = 6.4 \cdot 10^3$ J/kg °C as the specific heat of water at the operating coolant temperature yields

(f) $W = \dfrac{1}{c_p}\dfrac{PF_r}{(T_0|_{max} - T_i)} = \dfrac{1}{6.4 \cdot 10^3}\dfrac{3000 \cdot 10^6 \cdot 2.32}{(330 - 290)}$
$$= 27.2 \cdot 10^3\,\text{kg/s} = 27.2 \cdot 10^6\,\text{g/s}.$$

We determine the mean coolant velocity from $W = \rho A_{flow}\bar{v}$, where $A_{flow} = N \cdot (p^2 - \pi a^2)$, and we take the density of the pressurized water at 300 °C as 0.676 g/cm³. Thus

(g) $\bar{v} = \dfrac{W}{\rho N(p^2 - \pi a^2)} = \dfrac{27.2 \cdot 10^6}{0.676 \cdot 62{,}702 \cdot (1.536^2 - \pi 0.509^2)}$
$$= 415\,\text{cm/s} = 4.15\,\text{m/s}.$$

Finally, once the foregoing parameters are settled upon, the fuel enrichment and control poison requirements must be specified to obtain an acceptable value of k, the neutron multiplication.

The simplified modeling assumptions employed in this analysis have not taken into account a number of significant factors. For example, they do not allow space for control rods or structural supports, and these could add 10% or more to the core volume. Likewise, they do not include F_l, the local peaking factor, nor do they allow for substantial reductions in F_r that typically result from reducing the enrichment of the fuel elements or increasing the number of burnable poison rods to be placed in the highest flux regions of the core. The reduction in peaking factor in real designs substantially increases the average power density and decreases the core volume toward the PWR numbers quoted in Table 8.1. Nevertheless, the model demonstrates the substantial constraints under which the neutronics design of a power reactor core must proceed. Moreover, the foregoing calculations model represents only an attempt to obtain a workable set of core parameters; in reality, design is an iterative process. So, for example, if it turned out that the calculated coolant flow required a mean coolant velocity that is too large or resulted in an excessive pressure drop across the core, it would need to be reduced, and the other core parameters adjusted to accommodate the reduction.

8.4 Thermal Transients

As Eq. (8.32) indicates, the heat transferred from fuel to coolant under steady state conditions is just $P = (\overline{T}_f - \overline{T}_c)/R_f$. Conversely if cooling were cut off entirely, all of the heat produced would go into heating up the fuel adiabatically:

$$M_f c_f \frac{d}{dt} \overline{T}_f(t) = P(t), \qquad (8.47)$$

where M_f and c_f are the total fuel mass and specific heat, respectively. We may obtain an approximate lumped parameter model for thermal transients by combining these two expressions:

$$M_f c_f \frac{d}{dt} \overline{T}_f(t) = P(t) - \frac{1}{R_f}\left[\overline{T}_f(t) - \overline{T}_c(t)\right]. \qquad (8.48)$$

As justification for this equation consider the two bounding cases. At steady state the derivative on the left vanishes and Eq. (8.32) results. If all cooling is lost (say, by setting the convection from fuel to coolant to zero, which is equivalent to $R_f \to \infty$) the last term vanishes and

Eq. (8.47) results. A more convenient form of this model results from dividing by $M_f c_f$:

$$\frac{d}{dt}\bar{T}_f(t) = \frac{1}{M_f c_f}P(t) - \frac{1}{\tau}\left[\bar{T}_f(t) - \bar{T}_c(t)\right]. \qquad (8.49)$$

The first term on the right is the adiabatic heat-up rate, that is, the rate of fuel temperature rise if a power P is maintained but all cooling is cut off. The last term includes the core thermal time constant,

$$\tau = M_f c_f R_f, \qquad (8.50)$$

which is a measure of the time required for heat to be transferred from fuel to coolant.

The thermal time constant is useful for the analysis of transients; comparisons of it to other time constants such as the prompt and delayed neutron lifetimes, the rates of control rod insertions, and so on are often useful in judging the degree to which differing phenomena interact. For cores with liquid coolants, the value of τ is typically of the order of a few seconds, larger for oxide than for metal fuel. It is substantially larger for high temperature gas-cooled lattices, such as pictured in Fig. 4.1d, where the heat must pass through the graphite moderator before arriving at a coolant channel.

Fuel Temperature Transient Examples

Reactor power levels cannot change instantaneously, because of the kinetic effects of delayed neutrons discussed in Chapter 5. However, idealized step changes are useful in focusing attention on the significance of the thermal time constant. We consider two such hypothetical situations. In the first, a reactor operating at steady state power P_o is shut down instantaneously at $t = 0$. Equation (8.32) stipulates that the initial condition on the fuel temperature is $\bar{T}_f(0) = R_f P_o + \bar{T}_c$. Assuming the coolant temperature remains constant, Eq. (8.49) then reduces to

$$\frac{d}{dt}\bar{T}_f(t) = -\frac{1}{\tau}\left[\bar{T}_f(t) - \bar{T}_c\right], \quad t > 0. \qquad (8.51)$$

Application of the integrating factor technique described in Appendix A yields a solution of

$$\bar{T}_f(t) = \bar{T}_c + R_f P_o \exp(-t/\tau). \qquad (8.52)$$

Thus the fuel temperature decays exponentially, causing the fuel to lose half of its stored heat in a time $t = 0.693\tau$.

Next consider the converse situation in which the power suddenly jumps from zero to P_o. The initial condition will now be $\bar{T}_f(0) = \bar{T}_c$, and with constant coolant temperature, Eq. (8.49) takes the form

$$\frac{d}{dt}\bar{T}_f(t) = \frac{P_o}{M_f c_f} - \frac{1}{\tau}[\bar{T}_f(t) - \bar{T}_c], \quad t > 0. \tag{8.53}$$

This equation may be solved using the integrating factor, yielding a solution of

$$\bar{T}_f(t) = \bar{T}_c + R_f P_o[1 - \exp(-t/\tau)]. \tag{8.54}$$

At long times the exponential vanishes, causing $\bar{T}_f(\infty)$ to obey the steady state condition given by Eq. (8.32). At short times, measured as $t \ll \tau$, we may expand the exponential as $\exp(-t/\tau) \approx 1 - t/\tau$, which reduces the result to

$$\bar{T}_f(t) \approx \bar{T}_c + R_f P_o t/\tau = \bar{T}_c + \frac{P_o t}{M_f c_f}, \tag{8.55}$$

which is just the adiabatic heat-up rate. Thus we see that on time scales short compared to the thermal time constant, the core behaves adiabatically; for transients that are slow compared to the time constant, the core behaves in a quasi-steady state manner.

Coolant Temperature Transients

For simplicity, we have assumed that the coolant temperature remains constant in the above equations. This is often a reasonable approximation since if the coolant is a liquid the temperature changes in it are typically much smaller than in the fuel. Comparing the temperature drop between fuel and coolant, given by Eq. (8.32), with the temperature rise of the coolant, given by Eq. (8.37), provides a reference point:

$$\frac{\bar{T}_f - \bar{T}_c}{\bar{T}_o - T_i} = R_f W c_p. \tag{8.56}$$

Typically, for liquid-cooled reactors with oxide or carbide fuel $R_f W c_p \gg 1$.

For situations in which the time dependence of the average coolant temperature may become important, a differential equation approximating its behavior may be derived as follows. The power P appearing in Eqs. (8.32) and (8.40) represent, respectively, the heat flowing from the fuel to the coolant, and that being carried away by the coolant. Under steady state conditions they are equal. However, if

P in Eq. (8.32) is greater than its value in Eq. (8.40), the difference must appear as the rate at which internal energy is added to the coolant within the core. If M_c is the coolant mass within the core and c_p is its specific heat, then the rate of internal energy increase is

$$M_c c_p \frac{d}{dt} \bar{T}_c(t) = \frac{1}{R_f}\left[\bar{T}_f(t) - \bar{T}_c(t)\right] - 2W c_p\left[\bar{T}_c(t) - T_i\right]. \quad (8.57)$$

We may rewrite this equation in terms of two additional time constants

$$\frac{d}{dt} \bar{T}_c(t) = \frac{1}{\tau'}\left[\bar{T}_f(t) - \bar{T}_c(t)\right] - \frac{1}{\tau''}\left[\bar{T}_c(t) - T_i\right]. \quad (8.58)$$

Here

$$\tau' = \frac{M_c c_p}{M_f c_f}\tau, \quad (8.59)$$

but $\tau' \ll \tau$ because even for liquid coolants the heat capacity of the coolant in the core typically is much less than the fuel. The remaining time constant can be expressed in terms of t_c, the time required for the coolant to pass from core inlet to outlet. Suppose that A_{flow} is the flow area of the core, and ρ_c and $\bar{v}_c$ the density and average speed of the coolant; then $W = \rho_c A_{flow} \bar{v}_c$ and $M_c = \rho_c A_{flow} H$. Thus we have

$$\tau'' = \frac{M_c c_p}{2W c_p} = \frac{\rho_c A_{flow} H c_p}{2\rho_c A_{flow} \bar{v}_c c_p} = \frac{H}{2\bar{v}_c} = \frac{1}{2} t_c. \quad (8.60)$$

This time constant also tends to be substantially smaller than the fuel time constant. Thus the coolant follows the fuel surface transient quite rapidly such that the energy storage term on the left of Eq. (8.57) can be ignored in most cases.

To the extent that the foregoing assumptions hold, we can model the coolant by setting the left-hand side of Eq. (8.57) to zero. We then have

$$\bar{T}_c(t) = \frac{1}{1 + 2R_f W c_p}\left[2R_f W c_p T_i + T_f(t)\right]. \quad (8.61)$$

Combining this result with Eq. (8.49) then yields

$$\frac{d}{dt} \bar{T}_f(t) = \frac{1}{M_f c_f} P(t) - \frac{1}{\tilde{\tau}}\left[\bar{T}_f(t) - T_i\right], \quad (8.62)$$

where

$$\tilde{\tau} = \frac{2R_f W c_p}{1 + 2R_f W c_p} \tau. \tag{8.63}$$

However, because for most reactors $R_f W c_p \gg 1$, we may often make the approximations $\tilde{\tau} \approx \tau$ and

$$\bar{T}_c(t) \approx T_i + \frac{1}{2R_f W c_p} T_f(t). \tag{8.64}$$

Equations (8.62) and (8.64) provide a simple thermal model for use in analyzing reactor transients. In such transients thermal time constants—which are often of the order of a few seconds—frequently interact with the effects of the prompt and delayed neutron life times. Neutronic and thermal effects are strongly coupled through temperature-induced reactivity feedback. We take up these feedback effects in the next chapter.

Bibliography

Bonella, Charles F., *Nuclear Engineering*, McGraw-Hill, NY, 1957.

El-Wakil, M. M., *Nuclear Power Engineering*, McGraw-Hill, NY, 1962.

Glasstone, Samuel, and Alexander Sesonske, *Nuclear Reactor Engineering*, 3rd ed., Van Nostrand-Reinhold, NY, 1981.

Knief, Ronald A., *Nuclear Energy Technology: Theory and Practice of Commercial Nuclear Power*, McGraw-Hill, NY, 1981.

Lewis, E. E., *Nuclear Power Reactor Safety*, Wiley, NY, 1977.

Raskowsky, A., Ed., *Naval Reactors Physics Handbook*, U.S. Atomic Energy Commission, Washington, D.C., 1964.

Todreas, M. E., and M. S. Kazimi, *Nuclear Systems I & II*, Hemisphere, Washington, D.C., 1990.

Problems

8.1. The leakage probability of a power reactor is 0.08. As a first approximation to a new reactor an engineer estimates that the same power density can be achieved if the power is to be increased by 20%. Assuming the height to diameter ratio of the cylindrical core remains the same,

  a. What will the leakage probability be in the new reactor with the power increased by 20%?

  b. If k_∞ is proportional to the fuel enrichment, by what percent will the enrichment of the core need to be changed to accommodate the 20% increase in power?

8.2. A sodium-cooled fast reactor lattice is designed to have a migration length of 20 cm and a maximum power density of $500\,W/cm^3$. Three bare cylindrical cores with height to diameter ratios of one are to be built, with power ratings of 300 MW(t), 1000 MW(t), and 3000 MW(t). For each of the three cores determine the following:

a. The core height H.
b. The buckling B^2.
c. The nonleakage probability P_{NL}.

8.3. Consider a nonuniform cylindrical reactor with core radius R and height H. With control rods partially inserted, the power density distribution is approximated by

$$P'''(r,z) = A[1 - (r/R)^4][\cos(\pi z/H) - 0.25\sin(2\pi z/H)].$$

Assume the (r, z) origin is at the center of the reactor.

a. Find A in terms of the reactor power P.
b. Determine $f_r(r)$ and $f_z(z)$.
c. Determine F_r, F_z, and F_q (assuming $F_l = 1.1$).
d. Plot $f_r(r)$ and $f_z(z)$.

8.4. Consider a nonuniform cylindrical reactor with core radius R and height H. Two zones of fuel are employed, with a higher enrichment at the radial periphery to decrease the radial peaking factor. As a result, the power density is given by

$$P'''(r,z) = \begin{cases} A[1 - (r/R)^4]\cos(\pi z/H) & \text{for } 0 \leq r \leq \frac{3}{4}R \\ 1.7A[1 - (r/R)^4]\cos(\pi z/H) & \text{for } \frac{3}{4}R \leq r \leq R \end{cases}$$

Assume the (r, z) origin is at the center of the reactor.

a. Find A in terms of the reactor power P.
b. Determine $f_r(r)$ and $f_z(z)$.
c. Determine F_r and F_z.

8.5. Beginning with the heat balance $W_{ch}c_p dT_c(r,z) = q'(r,z)dz$ show that if the power distribution is axially symmetric, $q'(r,-z) = q'(r,z)$, then Eq. (8.39) for the average coolant temperature is exact.

8.6. You are to design a 3000 MW(t) pressurized water reactor. The reactor is a uniform bare cylinder with a height to diameter ratio of one. The coolant to fuel volume ratio is 2:1 in a square lattice. The volumes occupied by control and structural materials, as well as the extrapolation distances, can be neglected. The core inlet temperature is 290 °C. The reactor must operate under three thermal constraints: (1) maximum power density $= 250 \, \text{W/cm}^3$, (2) maximum cladding surface heat flux $= 125 \, \text{W/cm}^2$, and (3) maximum core outlet temperature $= 330$ °C. Determine the following:

a. The reactor dimensions and volume.
b. The fuel element diameter and lattice pitch.
c. The approximate number of fuel elements.
d. The mass flow rate and average coolant velocity.

8.7. A uniform cylindrical reactor core has a height to diameter ratio of one. The reactor is reflected both radially and axially with radial and axial reflector savings each being equal to M, the migration length of the core composition.

a. Show that the power peaking factor (with $F_l = 1.0$) is given by

$$F_q = \frac{1.889(1 + R/M)^{-2}(R/M)^2}{J_1[2.405(1 + R/M)^{-1}R/M] \sin[(\pi/2)(1 + R/M)^{-1}R/M]}.$$

b. Plot F_q vs R/M between $R/M = 5$ and $R/M = 50$ as well as the results for the same reactor without the reflector.

8.8.*Suppose the reflected reactor in problem 8.7 is a sodium-cooled fast reactor with $M = 18.0 \, \text{cm}$ and a power of 2000 MW(t). If the thermal design limits the maximum allowable power density to $450 \, \text{W/cm}^3$,

a. What is (1) the minimum value that the core radius can have, (2) the corresponding value of the core volume, and (3) the required value of k_∞ to maintain criticality?
b. Suppose that, to increase the thermal safety margins, it is decided to reduce the maximum permissible power density by 10%. What are the percentage changes for the reflected reactor in radius, volume, and k_∞? (Assume that M remains the same.)

8.9. Repeat problem 8.8 in the absence of the reflector.

8.10. Consider the PWR design at the end of Section 8.3. Suppose that by varying the enrichment in the fuel assemblies and distributing the control poisons in a nonuniform pattern the designers are able to reduce the radial and axial peaking factors to $F_r = 1.30$ and $F_z = 1.46$. Redesign the reactor by solving parts c through g of the pressurized water reactor example using those peaking factors.

8.11. An unachievable ideal would be a reactor with a perfectly flat flux distribution: $F_r = 1.00$ and $F_z = 1.00$. Repeat problem 8.10 for such an idealized reactor.

8.12. Suppose that the designers of the pressurized water reactor treated in Section 8.3 conclude that the thermal-hydraulic design must have larger safety margins by reducing the coolant flow velocity by 10% and the maximum coolant temperature by 5°C. The reactor physicists are asked to accommodate those changes by reducing the radial peaking factor. What percentage reduction would be required?

8.13. A reactor initially operating at a power P_o is put on a period T such that the power can be approximated as $P(t) = P_o \exp(t/T)$. Assuming that the coolant temperature is maintained at its initial value $T_c(0)$, solve Eq. (8.48) and show that the fuel temperature will be

$$T_f(t) = T_c(0) + \frac{P_o R_f}{1 + \tau/T} [\exp(t/T) + (\tau/T) \exp(-t/\tau)].$$

8.14. In the prismatic block form of the graphite-moderated gas-cooled reactor the heat passes through the moderator before reaching the coolant. Figures 4.1d and 4.2c show such a configuration. Develop a set of three coupled differential equations in forms similar to Eqs. (8.49) and (8.57) that describe the transient heat transfer in such a reactor. Assume that the heat transfer between fuel and moderator and between moderator and coolant is described by $P(t) = [\bar{T}_f(t) - \bar{T}_m(t)]/R_1$ and $P(t) = [\bar{T}_m(t) - \bar{T}_c(t)]/R_2$, respectively. Assume that W is the mass flow rate, and that the masses, specific heats, and densities of fuel, moderator, and coolant are given by M_i, c_i, and ρ_i with $i = f$, m, and c.

8.15.*A sodium-cooled fast reactor has the following characteristics

$$P = 2400 \, \text{MW(t)}, \qquad W = 14{,}000 \, \text{kg/s},$$
$$\tau = 4.0 \, \text{s}, \qquad c_p = 1250 \, \text{J/(kg °C)},$$
$$M_f c_f = 13.5 \times 10^6 \, \text{J/°C}, \qquad M_c c_p = 1.90 \times 10^6 \, \text{J/°C}.$$
$$T_i = 360 \, °\text{C},$$

Suppose the reactor undergoes a sudden trip, which may be approximated by setting the power equal to zero. Assuming the inlet temperature remains at its initial value, find the core outlet temperature and plot your results.

8.16. *Assume that the reactor in the preceding problem suffers a control failure and undergoes a power transient $P(t) = P(0)[1 + 0.25t]$, where t is in seconds. The shutdown system trips the reactor if the outlet temperature rises by more than $40\,°C$.

    a. Determine the coolant outlet temperature transient and plot your result until the temperature rises by $40\,°C$.

    b. At what time does the reactor shutdown system terminate the transient?

CHAPTER 9

Reactivity Feedback

9.1 Introduction

In Chapters 5 through 7, where the time and spatial distributions of neutrons in reactors are examined, the reader learned that with a specified set of cross sections the criticality equations allow reactors to operate at any power, without affecting the multiplication; the power level simply provides the normalization for the flux solution. The independence of power level from multiplication, however, only holds when the neutron flux is small. At higher power levels the temperatures of the fuel and coolant increase. Thermal expansion and other phenomena, such as Doppler broadening, then bring about changes in the cross sections, and concomitant changes in the value of k. In Chapter 8 we examined the temperature effects associated with power level changes. In this chapter we complete the loop by examining the reactivity feedback resulting from those changes in temperature.

We begin by defining reactivity coefficients and then look at those caused specifically by fuel and by moderator temperature changes. We proceed by examining the reactivity effects that take place under various operating conditions, such as heat up from room temperature to operating conditions, changes in power level, transients, and so on. After examining reactivity control and the concepts of excess reactivity, temperature, and power defects, we conclude with a discussion of reactor transients.

9.2 Reactivity Coefficients

The treatment of reactor kinetics in Chapter 5 indicates that the power produced by a reactor is determined by the time-dependent multiplication $k(t)$, or correspondingly the reactivity,

$$\rho(t) = \frac{k(t) - 1}{k(t)}. \tag{9.1}$$

221

In some situations it is possible to determine *a priori* the reactivity versus time. Such is often the case, for example, when control rods of known properties are moved in a prescribed manner in a reactor operating at very low neutron flux levels. As soon as a reactor produces enough power to raise the temperatures in the core above their ambient levels, however, the material densities as well as some of the microscopic cross sections become affected. Because the core multiplication depends on these densities, a reactivity feedback loop is established in which k, and thus ρ, is dependent on temperatures and densities that in turn are determined by the reactor power history. The understanding of the nature of mechanisms leading to reactivity feedback is essential to the analysis of power reactor behavior.

In relating the incremental reactivity changes to reactor multiplication, the following approximation is nearly universal:

$$d\rho = dk/k^2 \approx dk/k = d(\ln k). \qquad (9.2)$$

Because k rarely differs from one by more than a few percent, little error is introduced into ρ. Moreover, the logarithmic form of Eq. (9.2) facilitates the analysis of reactivity effects: Because k frequently is approximated with the four factor formula, writing $d\rho = d(\ln k)$ allows the factors to be transformed from multiplicative to additive form. In addition, since we make widespread use of Eq. (3.2) for expressing the multiplication as a product of infinite medium and leakage effects,

$$k = k_\infty P_{NL}, \qquad (9.3)$$

taking the logarithmic differential of k, and utilizing the definition of the nonleakage probability given by Eq. (7.49), yields additive terms for infinite medium and leakage effects:

$$\frac{dk}{k} = \frac{dk_\infty}{k_\infty} - \frac{M^2 B^2}{1 + M^2 B^2} \left(\frac{dM^2}{M^2} + \frac{dB^2}{B^2} \right). \qquad (9.4)$$

Moreover, since the leakage probability for large power reactors usually is quite small,

$$P_L = \frac{M^2 B^2}{1 + M^2 B^2} \ll 1. \qquad (9.5)$$

In large, loosely coupled cores changes in k_∞ often dominate, justifying the following approximation:

$$\frac{dk}{k} \approx \frac{dk_\infty}{k_\infty}. \qquad (9.6)$$

For thermal reactors Eq. (9.2) combined with the four factor formula, $k_\infty = \varepsilon p f \eta_T$, introduced in Section 4.4, allows further subdivision of reactivity effects into additive terms:

$$\frac{dk_\infty}{k_\infty} = \frac{d\varepsilon}{\varepsilon} + \frac{dp}{p} + \frac{df}{f} + \frac{d\eta_T}{\eta_T}. \qquad (9.7)$$

In what follows we examine a simplified model in which the multiplication changes only with the average fuel and coolant temperatures, $\bar{T}_f$ and $\bar{T}_c$:

$$k_\infty = k_\infty(\bar{T}_f, \bar{T}_c). \qquad (9.8)$$

We may then use partial derivatives to write

$$\frac{dk_\infty}{k_\infty} = \frac{1}{k_\infty}\frac{\partial k_\infty}{\partial \bar{T}_f}d\bar{T}_f + \frac{1}{k_\infty}\frac{\partial k_\infty}{\partial \bar{T}_c}d\bar{T}_c. \qquad (9.9)$$

Assuming the coolant and moderator are one and the same allows us to simply replace $\bar{T}_c$ by $\bar{T}_m$, and apply the four factor formula:

$$\frac{1}{k_\infty}\frac{\partial k_\infty}{\partial \bar{T}_x} = \frac{1}{\varepsilon}\frac{\partial \varepsilon}{\partial \bar{T}_x} + \frac{1}{p}\frac{\partial p}{\partial \bar{T}_x} + \frac{1}{\eta_T}\frac{\partial \eta_T}{\partial \bar{T}_x} + \frac{1}{f}\frac{\partial f}{\partial \bar{T}_x}, \qquad (9.10)$$

where $x = f$ and m indicate fuel and moderator, respectively. For reactors in which the moderator is distinct from the coolant, the four factor expressions introduced in Chapter 4 can be appropriately modified to take into account both moderator and coolant temperatures, $k_\infty(\bar{T}_f, \bar{T}_c, \bar{T}_m)$, and then we apply Eq. (9.10) to all three temperatures. If the coolant is a gas, however, it is usually adequate to consider only fuel and moderator temperature coefficients, since the density of the coolant gas is too small for its temperature to have an appreciable effect on reactivity.

Fuel Temperature Coefficient

Doppler broadening of the resonance capture cross sections of the fertile material accounts for the dominant part of the fuel temperature coefficient in low enrichment thermal power reactors and makes a substantial contribution in fast reactors as well. In thermal systems the effect appears as a decrease in the resonance escape probability with increased fuel temperature. There is no Doppler effect on ε, its contribution coming from energies well above where fuel resonance cross sections occur. There may be relatively minor changes in η_T and f in situations where substantial plutonium-239 is present, as it has a resonance only slightly above thermal energies. The latter effects are neglected in what follows, however, because they tend to be small compared to the change in the resonance escape probability.

The Doppler effect arises from the dependence of neutron cross sections on the relative speed between neutron and nucleus. The resonance cross sections are sharply peaked in energy, as shown, for example, in Fig. 4.6. For a given neutron speed the cross section must be averaged over the range of relative speeds resulting from the thermal motions of the fuel atoms, which constitute a Maxwell-Boltzmann distribution discussed in Chapter 2. This averaging has the net effect of slightly smearing the resonance in energy, causing them to appear wider and less peaked. The smearing becomes more pronounced as the fuel temperature rises, as shown in the exaggerated cross section curves in Fig. 4.6a. Thus we may approximate α_f, the fuel temperature coefficient, as

$$\alpha_f = \frac{1}{k}\frac{\partial k}{\partial \bar{T}_f} \approx \frac{1}{p}\frac{\partial \bar{p}}{\partial \bar{T}_f}. \tag{9.11}$$

In power reactor cores the concentration of resonance absorbers in the fuel is quite large. The effect is then to depress the neutron flux in the fuel both spatially, making $\varphi_f(E)/\varphi_m(E) < 1$, and in energy, as Fig. 4.6b indicates. The flux depression, known as resonance self-shielding, is most pronounced where the resonance cross section is largest. More rigorous analysis shows that as the fuel temperature increases, the flux depression or self-shielding becomes less pronounced, as sketched in Fig. 4.6b. The net effect is to increase the absorption rate and also the resonance integral. By inserting Eq. (4.40) for the resonance escape probability into Eq. (9.11) and performing the differentiation we obtain

$$\alpha_f = -\frac{V_f N_{fe}}{V_m \xi^m \Sigma_s^m}\frac{\partial I}{\partial \bar{T}_f} = -\ln\left(\frac{1}{p}\right)\frac{1}{I}\frac{\partial I}{\partial \bar{T}_f}. \tag{9.12}$$

An approximation for the temperature dependence of the resonance integral that is useful for thermal reactor analysis is given by

$$I \approx I(T_o)[1 + \tilde{\gamma}(\sqrt{\bar{T}_f} - \sqrt{T_o})], \tag{9.13}$$

where $\bar{T}_f$ and T_o are absolute temperatures in degrees kelvin, with the reference temperature taken as $T_o = 300\,\text{K}$. The coefficient is a function of the surface to mass ratio $(S/M = 4/\rho D)$ of a cylindrical fuel element of diameter D cm:

$$\tilde{\gamma} = C_1 + C_2(4/\rho D), \tag{9.14}$$

TABLE 9.1
Constants C_1 and C_2 for Eq. (9.14)

Fuel	C_1	C_2
Uranium metal	0.0048	0.0064
UO_2	0.0061	0.0047

Source: Pettus, W. G., and M. N. Baldwin, "Resonance Absorption in U^{238} Metal and Oxide Rods," Babcock and Wilcox Company Report No. BAW-1244, 1962.

where Table 9.1 displays the constants. Differentiating Eq. (9.13), we obtain

$$\frac{1}{I}\frac{\partial I}{\partial \bar{T}_f} = \frac{\tilde{\gamma}}{2\sqrt{\bar{T}_f}}\frac{I(T_o)}{I(\bar{T}_f)}. \qquad (9.15)$$

With $I(T_o)/I(\bar{T}_f) \approx /$, inserting this expression into Eq. (9.12) then yields

$$\alpha_f = -\frac{\tilde{\gamma}}{2\sqrt{\bar{T}_f}}\ln[1/p(T_o)]. \qquad (9.16)$$

Other effects may make smaller contributions to the fuel temperature coefficient. Fuel expansion, for example, causes perturbations in the four factors; however, these effects tend to be small in low enrichment reactors relative to that of Doppler broadening.

Moderator Temperature Coefficient

In those thermal reactors in which the moderator is a liquid, contributions to the moderator coefficient,

$$\alpha_m = \frac{1}{k}\frac{\partial k}{\partial \bar{T}_m}, \qquad (9.17)$$

derive primarily from density changes, with changes in the thermal neutron energy spectrum playing a secondary role. To illustrate, we first expand α_m in terms of N_m, the moderator atom density:

$$\alpha_m = \frac{1}{k_\infty}\frac{\partial k_\infty}{\partial N_m}\frac{\partial N_m}{\partial \bar{T}_m}. \qquad (9.18)$$

The atom density change in turn relates to the temperature coefficient through the volumetric coefficient of thermal expansion at constant pressure:

$$\beta_m = -\frac{1}{N_m}\frac{\partial N_m}{\partial \bar{T}_m}. \qquad (9.19)$$

Combining the two equations then yields

$$\alpha_m = -\beta_m N_m \frac{1}{k_\infty} \frac{\partial k_\infty}{\partial N_m}. \tag{9.20}$$

Consider next the dependence of the four factors on N_m. The fast fission factor increases somewhat with decreased moderator density because the neutrons are not slowed down as effectively to below the energy range where fast fission takes place. The effect is small, however, as is the effect on η_T, compared to that on the resonance escape probability and the thermal utilization. Thus we approximate

$$\frac{1}{k_\infty} \frac{\partial k_\infty}{\partial N_m} \simeq \frac{1}{p} \frac{\partial p}{\partial N_m} + \frac{1}{f} \frac{\partial f}{\partial N_m}. \tag{9.21}$$

Differentiating the expressions for p and f, given by Eqs. (4.54) and Eq. (4.55), with respect to N_m yields

$$\frac{1}{p} \frac{\partial p}{\partial N_m} = \frac{1}{N_m} \ln(1/p) \tag{9.22}$$

and

$$\frac{1}{f} \frac{\partial f}{\partial N_m} = -\frac{1}{N_m}(1 - f). \tag{9.23}$$

Combining Eqs. (9.20) through (9.23), we find

$$\frac{1}{k_\infty} \frac{\partial k_\infty}{\partial T_m} = -\beta_m[\ln(1/p) - (1 - f)]. \tag{9.24}$$

A decrease in moderator density decreases the effectiveness by which neutrons are slowed down through the resonance region. Hence the resonance absorption increases, causing the resonance escape probability to decrease. The lower moderator density, however, causes the thermal utilization to increase, resulting in a positive temperature effect from the second term in Eq. (9.24).

Although exceptions may occur under some conditions, generally in liquid-moderated reactors the decreasing moderator density is the dominant effect and causes the moderator temperature coefficient to be negative. Increases in moderator temperature also cause hardening of the thermal neutron spectrum, as indicted in Figure 3.5. In liquid-moderated reactors, the effects of spectral hardening are small compared to that of the reduced moderator density. In solid-moderated reactors, however, the thermal expansion has a much smaller effect, and spectral hardening dominates the determination of the temperature coefficient. The spectral effects depend on complication interactions to thermal neutron scattering in solid crystal structures and the non $1/v$ dependence of some isotopes' absorption cross sections. As a result moderator temperature coefficients in solid-moderated systems may be positive over some temperature ranges. In such situations reactor stability requires that the negative fuel temperature coefficient have a larger magnitude.

Fast Reactor Temperature Coefficients

The temperature coefficients of fast reactors differ in a number of respects from those of thermal reactors. First, the sizes of fast reactors measured in migration lengths tend to be substantially smaller than those for thermal reactors. This results both from the higher power density designs and from the fact that neutron cross sections are smaller for fast than for thermal spectrum neutrons. Consequently, the leakage terms in Eq. (9.4) have more substantial effects on reactivity.

In fast as well as in thermal reactors the Doppler broadening of capture resonances accounts for the major part of the negative fuel temperature coefficient. It is, however, smaller in magnitude because, as indicated by the spectra in Fig. 3.6, only a fraction of the neutrons in a fast reactor are slowed to the energy range where the large capture resonances in fertile materials occur.

Coolant temperature coefficients in fast reactors are yet more difficult to predict with elementary models, for they stem from the difference between two competing effects. With increased temperature, the coolant density decreases. Since in liquid-cooled fast reactors the coolant tends to degrade the neutron spectrum to lower energies, removal of coolant atoms hardens the neutron spectrum. As a result the larger value of $\eta(E)$ at higher neutron energies, as indicated in Fig. 3.2, causes the increased coolant temperature to increase the value of k_∞. Conversely, the decreased density that results from increased coolant temperature lengthens the migration length. Examination of Eq. (9.4) indicates that an increase in M will also increase the neutron leakage, decreasing the nonescape probability and decreasing the reactivity. Determining which of these effects is larger requires computational models that are beyond the level of this text.

9.3 Composite Coefficients

Chapter 8 provides a reactor model in terms of the average temperature of the fuel and coolant, $\bar{T}_f$ and $\bar{T}_c$. We can model the reactivity feedback by expanding it in terms of these temperatures:

$$d\rho_{fb} = \frac{1}{k}\frac{\partial k}{\partial \bar{T}_f}d\bar{T}_f + \frac{1}{k}\frac{\partial k}{\partial \bar{T}_c}d\bar{T}_c, \qquad (9.25)$$

where, for low leakage cores, the coefficients may be modeled solely in terms of k_∞ with little loss of accuracy. Reactivity coefficients couple to these temperatures in three distinctly different sets of circumstances. These lead to the definitions of the prompt and isothermal temperature coefficients, and to the power coefficient.

Prompt Coefficient

A large reactivity insertion will cause a reactor to go on such a short period that the power changes significantly over time spans that are short compared to the number of seconds needed to transfer heat from fuel to coolant. Over such short time spans the amount of heat transferred to the coolant is not enough to increase its temperature appreciably. As a result the dominant feedback reactivity comes directly from the heating of the fuel. Thus we take $d\bar{T}_c \approx 0$, reducing Eq. (9.25) to

$$d\rho_{fb} = \frac{1}{k}\frac{\partial k}{\partial \bar{\bar{T}}_f}d\bar{T}_f. \qquad (9.26)$$

The quantity

$$\alpha_f = \frac{d\rho_{fb}}{d\bar{T}_f} = \frac{1}{k}\frac{\partial k}{\partial \bar{\bar{T}}_f} \qquad (9.27)$$

is alternately known as the fuel temperature or the prompt coefficient, because it provides prompt feedback in the case of a sudden change in power. For a reactor to be stable, it must be negative.

A second mechanism does cause some instantaneous coolant heating. The increased number of neutrons colliding elastically with coolant atoms transfers some of their kinetic energy to the coolant atoms. However, this effect is significant compared to the Doppler broadening of the fertile material resonance only for systems fueled with highly enriched uranium—such as those found in naval propulsion reactors—where the presence of uranium-238 is greatly diminished.

Isothermal Temperature Coefficient

In many power reactors the entire core is brought very slowly from room temperature to the operating inlet coolant temperature, either by operating the reactor at a very low power or by using another source of heat such as that generated by the coolant pumps. In such an operation a reasonable approximation is to assume that the core behaves isothermally with a temperature $\bar{T}$, that is,

$$\bar{T}_f = \bar{T}_c = T_i = \bar{T}. \qquad (9.28)$$

Dividing Eq. (9.25) by $d\bar{T}$ yields the isothermal temperature coefficient,

$$\alpha_T \equiv \frac{d\rho_{fb}}{d\bar{T}} = \frac{1}{k}\frac{\partial k}{\partial \bar{T}_f} + \frac{1}{k}\frac{\partial k}{\partial \bar{T}_c}, \tag{9.29}$$

or equivalently $\alpha_T = \alpha_f + \alpha_c$, which accounts for the reactivity feedback under such circumstances.

Power Coefficient

Operating a reactor at more than a small percentage of its rated power causes the fuel temperature to rise significantly above that of the coolant, and the average coolant temperature to increase above that at the core inlet. Thus $\bar{T}_f > \bar{T}_c > T_i$, and we must replace the isothermal temperature coefficient with a reactivity coefficient that is applicable to a reactor operating at power. To obtain the power coefficient we divide Eq. (9.25) by an incremental power change, dP:

$$\alpha_P \equiv \frac{d\rho_{fb}}{dP} = \frac{1}{k}\frac{\partial k}{\partial \bar{T}_f}\frac{d\bar{T}_f}{dP} + \frac{1}{k}\frac{\partial k}{\partial \bar{T}_c}\frac{d\bar{T}_c}{dP}. \tag{9.30}$$

Provided the power changes are executed in a quasi-static manner, that is, the power changes are slow compared to the time required to remove heat from the fuel to the core outlet, the steady state heat transfer relationships derived in Chapter 8 are applicable. Differentiating Eqs. (8.41) and (8.40) with respect to power yields

$$\frac{d\bar{T}_f}{dP} = R_f + \frac{1}{2Wc_p} \tag{9.31}$$

and

$$\frac{d\bar{T}_c}{dP} = \frac{1}{2Wc_p}. \tag{9.32}$$

Inserting these expressions into Eq. (9.30) then expresses the power coefficient as

$$\alpha_P = \left(R_f + \frac{1}{2Wc_p}\right)\frac{1}{k}\frac{\partial k}{\partial \bar{T}_f} + \frac{1}{2Wc_p}\frac{1}{k}\frac{\partial k}{\partial \bar{T}_c}, \tag{9.33}$$

or, equivalently, as $\alpha_P = R_f\alpha_f + (2Wc_p)^{-1}(\alpha_f + \alpha_c)$.

Temperature and Power Defects

The magnitudes as well as the signs of the temperature and power coefficients strongly affect the control of reactivity in a power reactor. The temperature and power defect concepts make their influence clear.

Suppose we ask how much reactivity the control system must add to bring a reactor core from room temperature to the operating inlet temperature. The control reactivity must just compensate for the negative feedback reactivity. We find the reactivity decrease by integrating the isothermal temperature coefficient of Eq. (9.29) from room temperature, T_r, to the operating inlet temperature T_i:

$$D_T = \int_{T_r}^{T_i} \alpha_T(\bar{T})d\bar{T} \tag{9.34}$$

The temperature defect, as defined here, is negative provided the temperature coefficient is also negative. To compensate for it an equal amount of reactivity must be added to the reactor through withdrawal of control rods or other means in order to bring the reactor from room to operating temperature. These two states normally are referred to as cold critical and hot zero-power critical, respectively.

As the reactor power increases above the hot zero-power state to its rated level, a negative power coefficient causes the reactivity to decrease further by an amount determined by integrating the fuel and coolant temperature coefficients between inlet and full power conditions:

$$D_P = \int_{T_i}^{\bar{T}_f(p)} \alpha_f(\bar{T}_f)d\bar{T}_f + \int_{T_i}^{\bar{T}_c(p)} \alpha_e(\bar{T}_c)d\bar{T}_c. \tag{9.35}$$

where $\bar{T}(P) = \left[R_f + (2Wc_p)^{-1}\right]P + T_i$ and $\bar{T}_c(P) = (2Wc_p)^{-1}P + T_i$. This quantity is defined as the power defect. Because it is negative if the temperature coefficients are negative, the control system must add an equal amount of reactivity to bring the reactor from zero to full power. If the coefficients are temperature in dependent, then we have simply $D_P = \alpha_R D$.

9.4   Excess Reactivity and Shutdown Margin

For any well-defined state of a reactor core, we define the excess reactivity ρ_{ex} as the value that ρ would take if all movable control poisons were instantaneously eliminated from the core. Large excess reactivities are undesirable because they require large amounts of neutron poisons to be present in the core to compensate for them. The more control poison that is present in a core, the more extreme is the care that must be taken: The possibility of events that could lead to rapid ejection of a large enough fraction of that poison to bring the reactor near to prompt critical must be eliminated. Reactor designs must be shown to survive many hypothetical accidents without damage, including the ejection of a control rod, the uncontrolled withdrawal of a control rod bank, rapid dilution of soluble boron poison in water coolant, and so on. As a result, strict limitations are placed on the maximum amount of reactivity allowed in any one control rod, and also in any one bank of control rods. Thus the need for large excess reactivities complicates control system design by requiring increased numbers of control rod banks and/or more rods per bank. Such require-ment, in turn, increase the cost of the core since control rods occupy valuable space and add electromechanical complexity to the system.

Thus although negative temperature and power coefficients are necessary to ensure stability while the reactor is operating, inordi-nately large magnitudes of these quantities create problems by increasing the amount of excess reactivity that the control system must be able to overcome. The plots of excess reactivity in Fig. 9.1 versus time from beginning to end of core life (from BOL to EOL) illustrate this point by depicting the transitions between four states of the reactor: cold shutdown, cold critical, hot zero-power critical, and full power from the beginning through the end of core life. With this terminology, "cold" means ambient or room temperature, and "hot zero power" stipulates that the entire reactor has been heated to the temperature of the coolant inlet at full power, but essentially no power is being produced and thus the core is isothermal. As a reactor is taken from cold shutdown to full power, negative temperature

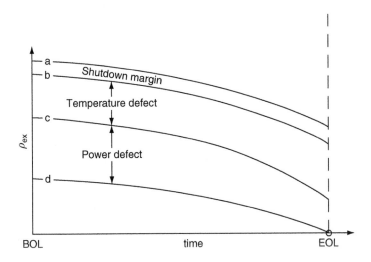

FIGURE 9.1   Excess reactivity vs time for different core states. (a) Cold shutdown, (b) cold critical, (c) hot zero power, (d) full power.

feedback causes excess reactivity to decrease, meaning that control poisons must be removed to keep the reactor critical. Conversely, when the reactor is shut down, temperatures decrease, causing excess reactivity to increase, and control poisons must be added.

Bringing the reactor critical from an initial shutdown state at room temperature decreases the excess reactivity from curve **a** to **b**. A minimum shutdown margin is invariably imposed by regulatory bodies and must be incorporated into control poison specifications. The core is then heated from cold critical to hot zero power, reducing the excess reactivity from curve **b** to **c**, with the difference being the magnitude of the temperature defect. Transitions between room temperature and hot zero-power temperature are carried out quite slowly, often limited by such mechanical considerations as excessive thermal stresses that would be induced in pressure vessels or piping if too rapid temperature transients were imposed. As the reactor is brought from zero to full power, more control poison must be withdrawn to compensate for the power defect in moving from curve **c** to **d**. Note that at full power, there is still excess reactivity in the core, and therefore control poison must be present. The shapes of the curves versus time are determined by fuel depletion and the buildup of fission products during core life. At the end of core life there is no excess reactivity at full power: All of the movable control poison has been extracted, and hence power must be reduced, or—more likely—the reactor is shut down for refueling. With online refueling, of course, curve **d** would be roughly horizontal, with very close to zero excess reactivity, and no specified end of core life.

In shutting down the reactor, we move back up the excess reactivity curve. The reactor may be made subcritical—or scrammed as the expression goes—from any power level. At that point all of the control poison is inserted. However, it must always be guaranteed that even as the reactor cools to ambient temperature, enough control poison remains to overcome the excess reactivity and maintain the shutdown margin.

9.5   Reactor Transients

The preceding section assumes implicitly that the transitions between room temperature, hot zero power, and full power take place very slowly. Such quasi-static transitions require the reactor to be on a period that is much longer than the time necessary for the heat produced in fission to be transferred to the coolant and convected out of the core. We may then employ steady state relationships for heat transfer as the reactor passes though a series of well-defined states, in which the excess reactivity is very close to being nullified by the control poisons. There are, however, situations that call for more rapid changes of neutron populations. In the startup of a reactor from cold critical, for example, the initial power is many orders of magnitude smaller than under operating conditions. Thus temperature feedback is insignificant, and a reactor period is chosen according to the kinetics equations of Chapter 5. At higher power levels where temperature feedback becomes important, changes in operating levels to follow power demand or for other reasons frequently must take place more rapidly than can be described with a quasi-static model. Finally, the examination of hypothetical reactor mishaps frequently involves transients that take place over time spans that are comparable to or shorter than those required to transfer heat from fuel to coolant.

Reactor Dynamics Model

To examine reactor transients we adopt a simple model in which a positive reactivity, $\rho_i(t)$, is inserted, and the feedback results from fuel and coolant temperature coefficients, α_f and α_c, respectively, which we assume are negative constants. If the reactor is critical and in steady state operation at the time of transient initialization, the total reactivity becomes

$$\rho(t) = \rho_i(t) - |\alpha_f| [\bar{T}_f(t) - \bar{T}_f(0)] - |\alpha_c| [\bar{T}_c(t) - \bar{T}_c(0)], \quad (9.36)$$

where we impose absolute value signs to indicate negative temperature coefficients. To analyze the transient we substitute this expression into the kinetics equations, given by Eqs. (5.47) and (5.48). Written in terms of the reactor power, and with the external source set equal to zero, they become

$$\frac{d}{dt} P(t) = \frac{[\rho(t) - \beta]}{\Lambda} P(t) + \sum_i \lambda_i \tilde{C}_i(t) \quad (9.37)$$

and

$$\frac{d}{dt} \tilde{C}_i(t) = \frac{\beta_i}{\Lambda} P(t) - \lambda_i \tilde{C}_i(t), \qquad i = 1, 2, 3, 4, 5, 6. \quad (9.38)$$

We approximate the temperatures using the transient heat transfer model given by Eqs. (8.62) and (8.64):

$$\frac{d}{dt} \bar{T}_f(t) = \frac{1}{M_f c_f} P(t) - \frac{1}{\tilde{\tau}} [\bar{T}_f(t) - T_i] \quad (9.39)$$

and

$$\bar{T}_c(t) = T_i + \frac{1}{2R_f W c_p} T_f(t). \quad (9.40)$$

Provided we approximate the thermal time constant as $\tilde{\tau} \approx \tau = M_f c_f R_f$, Eq. (9.40) allows the coolant temperature to be eliminated from Eq. (9.36):

$$\rho(t) = \rho_i(t) - |\alpha|\left[\bar{T}_f(t) - \bar{T}_f(0)\right], \tag{9.41}$$

where

$$|\alpha| = |\alpha_f| + \frac{1}{2R_f W c_p}|\alpha_c|. \tag{9.42}$$

If a reactor is critical before the transient is initiated, then $\tilde{C}_i(0) = (\beta_i/\lambda_i \Lambda)P(0)$ and the initial temperatures are given by the steady state conditions from Eqs. (8.40) and (8.41):

$$\bar{T}_f(0) = \left(R_f + \frac{1}{2W c_p}\right)P(0) + T_i \tag{9.43}$$

and

$$\bar{T}_c(0) = \frac{1}{2W c_p}P(0) + T_i. \tag{9.44}$$

Transient Analysis

We first apply the foregoing model to very slow transients, that is, ones in which the added reactivity increment $\rho_i(t) = \rho_o \ll \beta$ is so small that the resulting reactor period in the absence of feedback would be substantially longer than the fuel time constant, τ. In such situations, we may neglect the temperature derivative on the left of Eq. (9.39) and employ the resulting quasi-steady state heat transfer equation

$$\bar{T}_f(t) = \left(R_f + \frac{1}{2W c_p}\right)P(t) + T_i, \tag{9.45}$$

which when combined with Eq. (9.40) yields

$$\bar{T}_c(t) = \frac{1}{2W c_p}P(t) + T_i. \tag{9.46}$$

Substituting Eqs. (9.45) and (9.46) into Eq. (9.36) yields an approximate reactivity of

$$\rho(t) \approx \rho_o - \left[R_f|\alpha_f| + \frac{1}{2W c_p}(|\alpha_f| + |\alpha_c|)\right][P(t) - P(0)]. \tag{9.47}$$

Note, however, that the bracketed term is just equal to the magnitude of the power coefficient defined by Eq. (9.33). Hence,

$$\rho(t) \approx \rho_o - |\alpha_P|[P(t) - P(0)]. \tag{9.48}$$

In sufficiently slow transients, there will be a small initial prompt jump in the power, but then it will come to equilibrium as the rise in core temperatures compensates for ρ_o, sending $\rho(t) \to 0$. The new equilibrium power will be

$$P(\infty) = \frac{\rho_o}{|\alpha_P|} + P(0). \tag{9.49}$$

If small increments of reactivity are added successively at a very slow rate, then we may replace ρ_o with $\dot{\rho}t$, where $\dot{\rho}$ is referred to as the ramp rate, and obtain a reactor power that increases linearly with time:

$$P(t) = \frac{\dot{\rho}t}{|\alpha_P|} + P(0). \tag{9.50}$$

It cannot be stressed too strongly that the foregoing quasi-static analysis is only applicable to *very small* step reactivity insertions, or to exceedingly small ramp rates of reactivity addition. Otherwise, description of the transient requires the full set of Eqs. (9.36) through (9.42); Eqs. (9.49) and (9.50) only become applicable at long times, and only in those cases where equilibrium can be reestablished.

We now proceed to results that cannot be predicted with the simple quasi-static model. They require the full dynamics mode, Eqs. (9.36) through (9.42). Results obtained using parameters representative of a large pressurized water reactor appear in Figs. 9.2 and 9.3.

In Fig. 9.2 step reactivities of 0.20 and 0.95 dollars, that is, $\rho_o = 0.2\beta$ and $\rho_o = 0.95\beta$, are inserted into a reactor operating at full power. They demonstrate, respectively, what one might see during a normal operational transient, and a transient—approaching prompt critical—that is typical of those analyzed for hypothetical accidents. Note that curve **a**, resulting from the 0.20 dollar transient, undergoes a small prompt jump, followed by a slight gradual decrease to an equilibrium value determined by Eq. (9.49). In curve **b**, for the 0.95 dollar insertion, the reactor goes on to a very short period, and it exhibits a large power spike before the fuel gains enough temperature for negative feedback to negate the initial reactivity insertion. Over such short time spans heat does not have time to escape the fuel. Hence, at short times we may determine the fuel temperature by setting the last term in Eq. (9.39) to zero, and integrating to obtain

$$\bar{T}_f(t) \approx \frac{1}{M_f c_f} \int_0^t P(t')dt' + \bar{T}_f(0), \tag{9.51}$$

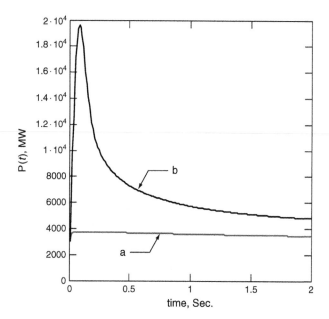

FIGURE 9.2 Power transients vs time initiated from full power for step reactivity insertions. (a) 0.20 dollars insertion, (b) 0.95 dollars insertion.

where $\bar{T}_c(t) \approx \bar{T}_c(0)$. Inserting this expression into Eq. (9.36) gives

$$\rho(t) \approx \rho_o - \frac{|\alpha_f|}{M_f c_f} \int_0^t P(t')dt', \tag{9.52}$$

where α_f is just equal to the prompt temperature coefficient given by Eq. (9.27). Thus over time spans that are short relative to the thermal time constant the feedback is proportional to the energy generated by the transient. Over the longer term the power will gravitate to a value determined by Eq. (9.49) only if the transient generates no boiling or other disruptions that invalidate the simple model presented here.

In practice, the rate at which reactivity is inserted often is as important as the total amount of reactivity available, whether that rate results from the planned withdrawal of a control rod bank or from a hypothesized accident scenario in which poison is suddenly ejected from the core. Indeed, because it is difficult to hypothesize physical mechanisms that could inject large amounts of reactivity "instantaneously" on time scales shorter than a prompt neutron lifetime, more realistic analysis normally centers on the rate at which reactivity might increase.

Figure 9.3 illustrates two power transients, both resulting from inserting reactivity at the very rapid rate of $\dot{\rho} = \beta$, that is 1.0 dollar/s, but with different initial conditions. We apply the same model, simply replacing ρ_o with $\dot{\rho}t$ in the foregoing equations. In the absence of

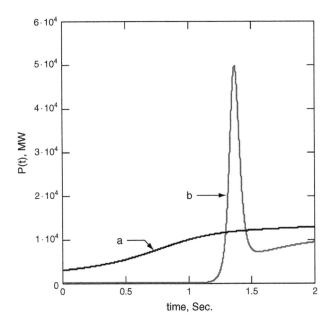

FIGURE 9.3 Power transients vs time for a reactivity insertion rate of 1.0 dollars/s. (a) Initiated from full power, (b) initiated from low power.

temperature feedback both transients would pass through prompt critical at one second. However, feedback does occur in both, and hence the severities of the transients depend strongly on the initial conditions. The differences become apparent by modifying short-time approximation given by Eq. (9.52) to apply to the ramp insertion of reactivity:

$$\rho(t) \approx \dot{\rho}t - \frac{|\alpha_f|}{M_f c_f} \int_0^t P(t')dt', \qquad (9.53)$$

which is reasonably applicable over the first one or two seconds, because we are using a fuel thermal time constant of $\tau = 4.5$ s. In curve **a**, the transient is initiated at full power, with $P(0) = 3000\,\text{MW(t)}$, allows negative feedback to accumulate rapidly. Thus, $\rho(t)$ never approaches prompt critical, and consequently the rate of power increase is quite modest. Curve **b**, however, results from initiating the transient at hot zero power, with $P(0) = 1.0\,\text{MW(t)}$. Hence the integral on the right of Eq. (9.53) builds at a much slower rate, allowing the reactivity to exceed prompt critical before being curtailed by negative reactivity feedback. A power spike thus ensues before sufficient energy is deposited in the fuel to bring the reactivity negative and terminate the transient.

The differences between curves **a** and **b** in Fig. 9.3 have strong implications for reactor operations. While at reactor is at full power, a reactor trip is normally initiated if the power rises by several percent. As Fig. 9.3 indicates, curve **a** reaches this point within a small fraction of a second. Thus if control rod banks are inserted within the next few seconds, the reactor is shut down following only a nominal rise in power. At lower powers, however, trip levels cannot be maintained too close to the initial power, for if they were, there would be no room to maneuver power to meet increased demand. Even if there were, curve **b** indicates that there is little possibility of generating a trip signal until well after one second into the transient. Because the power spike lasts substantially less than a second, there is no possibility that control rods can be inserted fast enough to limit its consequences. For this and a number of other reasons, the manipulation of reactors at very low power, where there is little or no negative temperature feedback, requires added care in order to eliminate the possibility of a so-called startup accident.

Bibliography

Bell, George I., and Samuel Glasstone, *Nuclear Reactor Theory*, Van Nostrand-Reinhold, NY, 1970.

Duderstadt, James J., and Louis J. Hamilton, *Nuclear Reactor Analysis*, Wiley, NY, 1976.

Hetrick, David, *Dynamics of Nuclear Systems*, American Nuclear Society, 1993.

Keepin, G. R., *Physics of Nuclear Kinetics*, Addison-Wesley, Reading, MA, 1965.

Lamarsh, John R., *Introduction to Nuclear Reactor Theory*, Addison-Wesley, Reading, MA, 1972.

Lewis, E. E., *Nuclear Power Reactor Safety*, Wiley, NY, 1977.

Ott, Karl O., and Robert J. Neuhold, *Introductory Nuclear Reactor Dynamics*, American Nuclear Society, 1985.

Schultz, M. A., *Control of Nuclear Reactors and Power Plants*, McGraw-Hill, NY, 1961.

Steward, H. B., and M. H. Merril, "Kinetics of Solid-Moderator Reactors," *Technology of Nuclear Reactor Safety*, T. J. Thompson and J. G. Beckerley, Eds., Vol. 1, MIT Press, Cambridge, MA, 1966.

Problems

9.1. Table 4.2 displays the four factors for a pressurized water reactor at 300 K. Assume the reactor is fueled with UO_2 fuel pins with diameters of 1.1 cm and a density of $11.0 \, g/cm^3$.

    a. Determine the fuel temperature coefficient and plot its value between 300 and 1000 K.

    b. Estimate the coolant temperature coefficient, assuming that the water has a coefficient of thermal expansion given by $\beta_m = 0.004 \, K^{-1}$.

9.2. Consider a hypothetical reactor in which all of the materials have the same volumetric coefficient of thermal expansion. Thus all of the nuclide densities decrease according to the same ratio: $N'/N = const. < 1$.

a. Show that the expansion with increased temperature has no effect on k_∞.
b. Using the facts that the core mass, NV, remains constant, and that $M \propto N^{-1}$, show from Eq. (9.4) that the reactivity change from expansion is negative, with a value of $\dfrac{dk}{k} = -\dfrac{4}{3} P_L \dfrac{dV}{V}$.

9.3. Assume a pressurized water reactor has the parameters specified in the example at the end of Section 8.3. Assume the core has a thermal resistance of $R_f = 0.50\,°C/MW(t)$. If the reactor fuel and moderator temperature coefficients are $\alpha_f = -3.2 \cdot 10^{-5}\,°C^{-1}$ and $\alpha_m = -1.4 \cdot 10^{-5}\,°C^{-1}$,

a. Determine the isothermal temperature coefficient.
b. Determine the power coefficient.

9.4. At full power a 1000 MW(t) sodium-cooled fast reactor has coolant inlet and outlet temperatures of 350 and 500 °C and an average fuel temperature of 1150 °C. The fuel and coolant temperature coefficients are $\alpha_f = -1.8 \times 10^{-5}/°C$ and $\alpha_c = +0.45 \times 10^{-5}/°C$.

a. Estimate the core thermal resistance and the mass flow rate, taking for sodium $c_p = 1250\,J/kg\,°C$.
b. Estimate the temperature and power defects, assuming a "cold" temperature of 180 °C.

9.5.*A 3000 MW(t) pressurized water reactor has the following specifications: core thermal resistance 0.45 °C/MW(t), coolant flow 68×10^6 kg/hr, coolant specific heat 6.4×10^3 J/kg °C. The fuel temperature coefficient is

$$\frac{1}{k}\frac{\partial k}{\partial \bar{T}_f} = -\frac{7.2 \cdot 10^{-4}}{\sqrt{273 + \bar{T}_f}}\,(°C)^{-1}$$

and the coolant temperature coefficient is

$$\frac{1}{k}\frac{\partial k}{\partial \bar{T}_c} = \left(30 + 1.5\bar{T}_c - 0.010\bar{T}_c^2\right) \cdot 10^{-6}\,(°C)^{-1}.$$

a. Over what temperature range is the core overmoderated?
b. What is the value of the temperature defect? Assume room temperature of 21°C and an operating coolant inlet temperature of 290 °C.
c. What is the value of the power defect?

9.6. A sodium-cooled fast reactor lattice is designed to have the following properties: migration length 18.0 cm and a maximum power density 450 W/cm³. Fractional sodium voiding results in the following reactivity effects:

$$\Delta k_\infty/k_\infty = +0.002, \quad \Delta M/M = +0.01.$$

Three bare cylindrical cores with height to diameter ratios of one are to be built with power ratings of 300 MW(t), 1000 MW(t), and 3000 MW(t).

a. Find H, the core height, B^2, the buckling, and k_∞ for each of the cores.
b. For each of the three cores, determine the reactivity change caused by the voiding.
c. Briefly interpret your results from part b.

9.7. For the reactor specified in problem 9.4 the power is maintained at 1000 MW(t) while the following quasi-static changes are made:

a. The inlet temperature is slowly decreased by 10 °C.
b. The flow rate is slowly increased by 10%.

For each of the cases determine by how much the reactivity must be increased or decreased to keep the reactor running at constant power.

9.8. In the prismatic block form of the graphite-moderated gas-cooled reactor the heat passes through the moderator before reaching the coolant. Figure 4.1d shows such a configuration. Assume that R_1 and R_2 are the thermal resistances between fuel and moderator and between moderator and coolant, respectively, and that W and c_p are the coolant mass flow rate and specific heat.

a. Develop a set of three coupled equations similar to those in Eqs. (8.40) and (8.41) to model the steady state heat transfer.
b. Determine the isothermal temperature coefficient in terms of the fuel, moderator, and coolant temperature coefficients.
c. Determine the power coefficient in terms of the same temperature coefficients.

9.9. Suppose a power reactor has negative values of α_f and α_c, the fuel and coolant temperature coefficients. Using the thermal hydraulic model developed in Chapter 8,

a. Show that if very slow changes take place in the coolant inlet temperature and mass flow rate, but no control poisons are added or subtracted, the power will undergo a quasi-static change of

$$dP = \frac{|\alpha_f + \alpha_c|\left(\dfrac{P}{2W^2 c_p}dW - dT_i\right)}{\left|\left(R_f + \dfrac{1}{2Wc_p}\right)\alpha_f + \dfrac{1}{2Wc_p}\alpha_c\right|}.$$

b. If the flow rate increases, does the power increase or decrease? Why?

c. If the inlet temperature increases, does the power increase or decrease? Why?

9.10.*Apply appropriate software to Eqs. (9.37) through (9.42) for a uranium-fueled reactor. Use the following parameters, which are typical of a large pressurized water reactor: $\Lambda = 50 \times 10^{-6}$ s, $\alpha = -4.2 \times 10^{-5}/°C$, $M_f c_f = 32 \times 10^6 J/°C$, and $\tau = 4.5$ s, and $R_f W c_p \gg 1$. With an initial steady state power of 10 MW, make a plot of the power versus time for the following:

a. A step reactivity increase of 10 cents.
b. A step reactivity increase of 20 cents.
c. A step reactivity decrease of 10 cents.
d. A step reactivity decrease of 20 cents.

9.11.*Using the data and initial conditions from problem 9.10 and applying appropriate software,

a. Determine what ramp rate of reactivity insertion will cause a power spike with a peak power that exceeds 100 MW(t).
b. Determine what ramp rate of reactivity insertion will cause a power spike with a peak that exceeds 1000 MW(t).

9.12.*Repeat problem 9.10 for a plutonium-fueled sodium-cooled fast reactor with the following parameters: $\Lambda = 0.5 \times 10^{-6}$s, $\alpha = -1.8 \times 10^{-5}/°C$, $M_f c_f = 5.0 \times 10^6 J/°C$, and $\tau = 4.0$ s.

9.13.*A 2400 MW(t) plutonium sodium-cooled fast reactor has the following characteristics:

$W = 14,000$ kg/s, $\alpha_f = -1.8 \times 10^{-5}/°C$,
$M_f c_f = 13.5 \times 10^6 J/°C$, $c_p = 1250 J/(kg°C)$,
$\Lambda = 0.5 \times 10^{-6}$ s, $T_i = 360°C$,
$\tau = 4.0$ s, $\alpha_c = +0.45 \times 10^{-5}/°C$.
$M_c c_p = 1.90 \times 10^6 J/°C$,

The reactor undergoes a loss of flow transient with $W(t) = W(0)/(1 + t/t_o)$, where $t_o = 5.0$ s. Employ appropriate software to Eqs. (9.36) through (9.40) to analyze the transient: Make plots of the reactor power, fuel, and coolant outlet temperatures for $0 < t < 20$ s. (Hint: Note that $\tilde{\tau}$ cannot be approximated by τ in this problem.)

CHAPTER 10

Long-Term Core Behavior

10.1 Introduction

This final chapter addresses the longer term changes that occur over the lifetime of a power reactor core. These are most directly tied to the evolution of the fuel composition and its by-products as a function of time. They fall into three categories: (1) the buildup and decay of radioactive products of fission, (2) fuel depletion, and (3) the buildup of actinides resulting from neutron capture in fissile and fertile materials. For the most part these phenomena take place on time scales that are substantially longer than those detailed thus far. Whereas fission products have half-lives from seconds to decades, many of the more important, such as xenon and samarium, have half-lives of several hours or more. The effects of fuel depletion are measured on yet longer time scales, typically weeks, months, or years. These times are much longer than those dealt with in earlier chapters. In reactor kinetics we dealt with prompt neutron lifetimes that are small fractions of a second and delayed neutron lifetimes of minutes of less. In energy transport we most typically dealt with thermal time constants ranging from seconds to minutes. These differences in time scale often serve to simplify the analysis. For example, in reactor kinetics we need not consider fuel depletion; it is much too slow a process. Conversely, in fuel management studies, we can assume reactor kinetics effects to be instantaneous and also neglect the effects of thermal transients.

10.2 Reactivity Control

As a power reactor operates the multiplication decreases with time as the fuel is depleted and fission products accumulate. To analyze these effects we model a thermal reactor utilizing the four factor formula, given by Eq. (4.24), multiplied by the nonleakage probability to account for the core's finite size:

$$k = \eta_T f \varepsilon p P_{NL}. \tag{10.1}$$

The effects of fuel burnup and fission product buildup appear primarily in the thermal cross sections, and therefore affect η_T and f. Utilizing Eqs. (4.48) and (4.49) we may write these more explicitly to obtain

$$k = \frac{\nu \Sigma_f^f}{\Sigma_a^f + \varsigma(V_m/V_f)\Sigma_\gamma^m} \varepsilon p P_{NL}, \qquad (10.2)$$

where we hereafter drop the subscript T that indicates thermal cross sections.

Fuel depletion causes the fission cross section to become time dependent, $\Sigma_f^f \rightarrow \Sigma_f^f(t)$. Fuel depletion also causes the fuel absorption cross sections to become time dependent, but in addition fission products generated in the fuel add to the capture cross section. Thus $\Sigma_a^f \rightarrow \Sigma_a^f(t) + \Sigma_a^{fp}(t)$. While fuel burns and fission products accumulate, of course, the reactor must be kept critical, that is, with $k = 1$. To accomplish this the control rods or other neutron poisons must be present at the beginning of core life; these are then extracted to maintain criticality as power is produced. These poisons appear as a control capture cross section $\Sigma_\gamma^{con}(t)$ added to the denominator of Eq. (10.2). Thus as detailed in Chapter 9, to maintain criticality we must have

$$1 = \frac{\nu \Sigma_f^f(t)}{\Sigma_a^f(t) + \Sigma_\gamma^{fp}(t) + \varsigma(V_m/V_f)\Sigma_\gamma^m + \Sigma_\gamma^{con}(t)} \varepsilon p P_{NL}. \qquad (10.3)$$

Near universal practice, however, is to calculate the multiplication with all of the removable control poisons withdrawn from the core:

$$k(t) = \frac{\nu \Sigma_f^f(t)}{\Sigma_a^f(t) + \Sigma_\gamma^{fp}(t) + \varsigma(V_m/V_f)\Sigma_\gamma^m} \varepsilon p P_{NL}. \qquad (10.4)$$

The excess reactivity is then defined as

$$\rho_{ex}(t) = \frac{k(t) - 1}{k(t)} \approx k(t) - 1, \qquad (10.5)$$

where $k(t)$ remains greater than one until the end of core life, at which time all of the movable control poison has been extracted.

Substantial amounts of excess reactivity at the time of reactor startup allow for extended core life before refueling must take place. As indicated in Chapter 9, however, large excess reactivities create challenges in the design of a reactor's control system. Control rods are the most common means for compensating for excess reactivity. However, in large, neutronically loosely coupled cores great care must be taken to ensure that their presence does not distort the flux distribution to the extent that excessive power peaking results. In pressurized water reactors dissolving a soluble neutron absorber in the coolant and varying the concentration with time serves to compensate for excess reactivity. Burnable poisons embedded in the fuel or other core constituents offer an additional means of limiting excess reactivity as well as mitigating localized power peaking.

Both fission product buildup and fuel depletion contribute to the deterioration of reactivity, but on different time scales. Fission products tend to build to a saturation value and then remain at constant concentration as indicated in Section 1.7. The two fission products with the largest capture cross section are xenon-135 and samarium-149, and these have half-lives measured in hours; they therefore come to equilibrium within days following startup or shut down. Fuel depletion evolves more slowly over time spans typically measured in weeks or months. Thus to great extent we can uncouple the two phenomena and treat them separately. We treat fission products first, and then fuel depletion and the buildup of transuranic nuclei; we conclude with a brief treatment of burnable poisons.

Before proceeding we note that we may also analyze fast reactors with a form similar to Eq. (10.4):

$$k(t) = \frac{\nu \Sigma_f^f(t)}{\Sigma_a^f(t) + (V_c/V_f)\Sigma_\gamma^c} P_{NL}. \qquad (10.6)$$

Here the factors ε, p, and ς are removed, and the cross sections are averaged over all neutron energies, instead of only the thermal energy range. We have also replaced m with c so the equation is applicable to fast as well as thermal reactors. The effects of the fission product buildup, however, are much less significant in fast reactors since they have substantial capture cross sections only for thermal neutrons.

10.3 Fission Product Buildup and Decay

As Chapter 1 details, fission reactions on average produce two radioactive fission fragments, each undergoing one or more subsequent radioactive decays. Many of the resulting fission products have measurable thermal absorption cross sections. Thus their buildup creates neutron poisons. The poisons' significance depends both on the fission product production and decay rates. For as we saw in Chapter 1 equilibrium is eventually reached if a radioisotope is produced at a constant rate, for after several half-lives the rate of decay will equal the rate of production. In tracking fission products in an operating reactor, however, we must also take into account the rate of destruction that results from neutron absorption.

Suppose we let $N(t)$ be the concentration of a particular fission product isotope at a time t following reactor startup, and specify the fission rate as $\bar{\Sigma}_f \phi$. If some fraction γ_{fp} of the fissions results in the production of the fission product, then the rate of production will be $\gamma_{fp} \bar{\Sigma}_f \phi$. The fission product will decay away at a rate $\lambda N(t)$, where λ is the isotope's decay rate. In addition, if the isotope has a thermal neutron absorption cross section of σ_a, it will undergo destruction at a rate of $\sigma_a N(t) \phi$. Thus its inventory will grow at a net rate of

$$\frac{d}{dt} N(t) = \gamma_{fp} \bar{\Sigma}_f \phi - \lambda N(t) - \sigma_a N(t) \phi. \tag{10.7}$$

Note that both of the loss terms are proportional to $N(t)$. Thus we may write this equation in the compacted form

$$\frac{d}{dt} N(t) = \gamma_{fp} \bar{\Sigma}_f \phi - \lambda' N(t), \tag{10.8}$$

where

$$\lambda' = \lambda + \sigma_a \phi, \tag{10.9}$$

and refer to the related quantity $t'_{1/2} = 0.693 / \lambda'$ as the effective half-life.

Since Eq. (10.8) has the same form as Eq. (1.39), we employ the same integrating factor technique to obtain

$$N(t) = \frac{\gamma_{fp} \bar{\Sigma}_f \phi}{\lambda'} [1 - \exp(-\lambda' t)]. \tag{10.10}$$

The concentration of each fission product depends strongly on the time t that the reactor has been operating, relative to the values of λ' and $\sigma_a \phi$. If $\lambda' t \ll 1$ then the concentration will increase linearly with time: $N(t) \approx \gamma_{fp} \bar{\Sigma}_f \phi t$. However at longer times, when $\lambda' t \gg 1$, the concentration will reach a saturation value of $\gamma_{fp} \bar{\Sigma}_f \phi / \lambda'$, and increase no further.

Note that a reactor's inventory of those isotopes that have reached saturation is proportional to $\bar{\Sigma}_f \phi$ and therefore to the power at which the reactor is operating, whereas the inventory of those for which $\lambda' t \ll 1$ is proportional to $\bar{\Sigma}_f \phi t$ and therefore to the total energy that has been produced by the reactor. Thus iodine-131 quickly reaches its saturation value since it has a half-life of 8.0 days; its inventory thereafter will be proportional to the reactor power. In contrast, cesium-137, with a half-life of 30.2 years, increases linearly with time over the few years that fuel is in the reactor, and its inventory is proportional to the total energy that has been produced from the fuel.

Xenon Poisoning

The most substantial fission product effects on reactor operation come from the exceedingly large absorption cross section of the isotope xenon-135, which measures 2.65×10^6 b. Xenon is produced both directly and from the decay of other fission products, most importantly iodine-135. The sequence of beta decays leading to xenon may be summarized as

$$
\begin{array}{ccccccc}
\textit{fission},\gamma_T & & \textit{fission},\gamma_I & & \textit{fission},\gamma_X & & \\
\downarrow & & \downarrow & & \downarrow & & \\
{}^{135}_{52}\text{Te} & \xrightarrow[11s]{\beta^-} & {}^{135}_{53}\text{I} & \xrightarrow[6.7hr]{\beta^-} & {}^{135}_{54}\text{Xe} & \xrightarrow[9.2hr]{\beta^-} {}^{135}_{55}\text{Cs} & \xrightarrow[2.3\times10^6yr]{\beta^-} {}^{135}_{56}\text{Ba,}
\end{array}
$$

$$(10.11)$$

where we have included the half-lives. In dealing with time spans measured in hours, we may assume the tellurium-135 decay is instantaneous and combine the yields of tellurium and iodine isotopes into γ_I; Table 10.1 gives values of γ_I and γ_X. Likewise, we may neglect the decay of cesium since its half-life is more than a million years.

With these simplifications, we need just two rate equations. Let I and X denote the concentrations of the iodine and xenon isotopes. We then have

$$\frac{d}{dt}I(t) = \gamma_I \bar{\Sigma}_f \phi - \lambda_I I(t) \tag{10.12}$$

and

$$\frac{d}{dt}X(t) = \gamma_X \bar{\Sigma}_f \phi + \lambda_I I(t) - \lambda_X X(t) - \sigma_{aX} X(t)\phi. \tag{10.13}$$

No neutron absorption term, $\sigma_{aI}I(t)\phi$, appears in the first equation since even at high flux levels iodine absorption is insignificant compared to its decay.

TABLE 10.1
Thermal Fission Product Yields in Atoms per Fission

Isotope	Uranium-235	Plutonium-239	Uranium-233
^{135}I	0.0639	0.0604	0.0475
^{135}Xe	0.00237	0.0105	0.0107
^{149}Pm	0.01071	0.0121	0.00795

Source: M. E. Meek and B. F. Rider, "Compilation of Fission Product Yields," General Electric Company Report NEDO-12154, 1972.

Following reactor start-up both iodine and xenon concentrations build from zero to equilibrium values over a period of several half-lives. Since the half-lives are in hours, after a few days equilibrium is achieved. The equilibrium concentrations result from setting the derivatives on the left of Eqs. (10.12) and (10.13) to zero:

$$I(\infty) = \gamma_I \bar{\Sigma}_f \phi / \lambda_I \tag{10.14}$$

and

$$X(\infty) = \frac{(\gamma_I + \gamma_X)\bar{\Sigma}_f \phi}{\lambda_X + \sigma_{aX}\phi}. \tag{10.15}$$

Note that for very high flux levels, where $\sigma_{aX}\phi \gg \lambda_X$, neutron absorption dominates over radioactive decay, and the maximum xenon concentration possible is $(\gamma_I + \gamma_X)\bar{\Sigma}_f / \sigma_{aX}$.

Next consider what happens following reactor shutdown. Let I_o and X_o be the concentrations of iodine and xenon at the time of shutdown. If the reactor is put on a large negative period, to first approximation we may assume that the shutdown is instantaneous compared to the time spans of hours over which the iodine and xenon concentrations evolve. This assumption allows us to determine the isotopes' concentrations by setting $\phi = 0$ in Eqs. (10.12) and (10.13). The solution of Eq. (10.12) yields the exponential decay of the iodine,

$$I(t) = I_o \exp(-\lambda_I t), \tag{10.16}$$

where t is now taken as the time elapsed since shutdown. Inserting this expression into Eq. (10.13), with $\phi = 0$, yields

$$\frac{d}{dt}X(t) = \lambda_I I_o \exp(-\lambda_I t) - \lambda_X X(t). \tag{10.17}$$

Employing the integrating factor technique detailed in Appendix A we obtain a solution of

$$X(t) = X_o e^{-\lambda_X t} + \frac{\lambda_I}{\lambda_I - \lambda_X} I_o (e^{-\lambda_X t} - e^{-\lambda_I t}). \tag{10.18}$$

The first term results from the decay of xenon. The second arises from the production of xenon—caused by iodine decay—following shutdown and then the subsequent decay of that xenon. If the reactor has been running for several days—long enough for iodine and xenon to reach equilibrium—we may replace I_o and X_o by the values of $I(\infty)$ and $X(\infty)$ given by Eqs. (10.14) and (10.15):

$$X(t) = \bar{\Sigma}_f \phi \left[\frac{(\gamma_I + \gamma_X)}{\lambda_X + \sigma_{aX}\phi} e^{-\lambda_X t} + \frac{\gamma_I}{\lambda_I - \lambda_X} \left(e^{-\lambda_X t} - e^{-\lambda_I t} \right) \right]. \qquad (10.19)$$

The effect on reactivity appears as a contribution of $\sigma_{aX}X(t)$ to the fission product term in the denominator of Eq. (10.4). It causes a negative reactivity that to a first approximation is proportional to $\sigma_{aX}X(t)$. Figure 10.1 contains plots of the negative reactivity versus time for a representative thermal reactor that has operated at four different flux levels. The curves provide insight into the challenges caused by xenon following shutdown. For a sufficiently large operating flux the xenon concentration actually rises following shutdown; for uranium-fueled reactors that flux is approximately 4×10^{11} n/cm^2/s. The peak concentration occurs at 11.3 hours after shutdown. Since xenon has a large absorption cross section, this

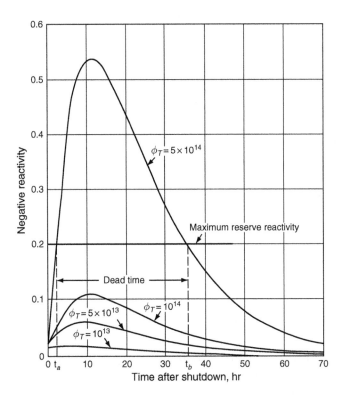

FIGURE 10.1 Xenon-135 reactivity transient following shutdown from four different flux levels (adapted from *Introduction to Nuclear Reactor Theory*, 1972, by John R. Lamarsh, Copyright by the American Nuclear Society, La Grange Park, IL).

increase causes a loss in reactivity that must be overcome if the reactor is to be restarted. Figure 10.1 also depicts a dead time over which the reactor could not be restarted if 0.2 is the reserve of excess reactivity at room temperature provided by the control system.

Xenon poisoning can cause difficulties at other times. If a reactor is operated in a periodic manner, for example, at lower power at night than in the daytime to follow electrical load demand, the time lag in the xenon buildup and decay must be compensated by the use of periodic insertion of control poisons. In large, neutronically loosely coupled cores, where control rod effects are localized, xenon may give rise to space–time oscillations. If these are not adequately damped, they may result in increased power peaking in local areas of the core and violate thermal limitations on local power density.

Samarium Poisoning

A second fission product with a large thermal absorption cross section is samarium-149. Although smaller than xenon, its cross section of 41,000 b must be taken explicitly into account. Samarium is stable, arising from the fission product chain

$$\overset{\text{fission, } \gamma_N}{\downarrow}$$
$$^{149}_{60}\text{Nd} \xrightarrow[1.7hr]{\beta^-} {}^{149}_{61}\text{Pm} \xrightarrow[53hr]{\beta^-} {}^{149}_{62}\text{Sm}. \tag{10.20}$$

Since the half-life of neodymium is much shorter than that of promethium, to a good approximation we can assume that the promethium is produced directly by fission; Table 10.1 gives the value of γ_P. Thus

$$\frac{d}{dt}P(t) = \gamma_P \bar{\Sigma}_f \phi - \lambda_P P(t) \tag{10.21}$$

and

$$\frac{d}{dt}S(t) = \lambda_P P(t) - \sigma_{aS} S(t)\phi. \tag{10.22}$$

The saturation activities reached after several half-lives of reactor operations result from setting the derivatives on the left of these equations to zero: $P(\infty) = \gamma_P \bar{\Sigma}_f \phi / \lambda_P$ and $S(\infty) = \gamma_P \bar{\Sigma}_f / \sigma_{aS}$. The solutions of Eqs. (10.21) and (10.22) following shutdown yield

$$P(t) = P_o \exp(-\lambda_P t) \tag{10.23}$$

and

$$S(t) = S_o + P_o\left(1 - e^{-\lambda_P t}\right),\qquad(10.24)$$

where t is the time elapsed since shutdown. Provided the saturation values have been reached by the time of shutdown, we can combine equations to obtain

$$S(t) = \frac{\gamma_P \bar{\Sigma}_f}{\sigma_{aS}} + \frac{\gamma_P \bar{\Sigma}_f \phi}{\lambda_P}\left(1 - e^{-\lambda_P t}\right).\qquad(10.25)$$

Analogous to xenon poisoning, samarium causes a reactivity loss that is approximately proportional to $\sigma_{aS}S(t)$. Figure 10.2 illustrates the reactivity loss for the same set of reactor parameters used in Fig. 10.1. The samarium concentration rises following shutdown, and at long times is greater than the saturation value by an amount

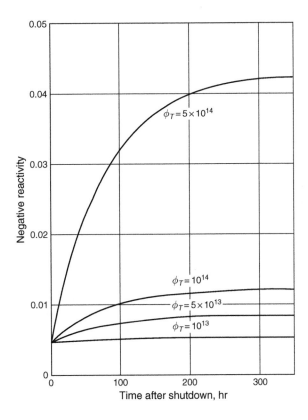

FIGURE 10.2 Sumarium-149 reactivity transient following shutdown from four different flux levels (adapted from *Introduction to Nuclear Reactor Theory*, 1972, by John R. Lamarsh, Copyright by the American Nuclear Society, La Grange Park, IL).

$\gamma_P \bar{\Sigma}_f \phi / \lambda_P$. Thus to restart the reactor after an extended shutdown sufficient additional reactivity must be available to overcome the added samarium that resulted from the promethium decay.

10.4   Fuel Depletion

Heretofore we have written the power density simply as $P''' = \gamma \Sigma_f \phi$. To examine the long-term behavior of a power reactor core, however, we must divide this term into its component parts. Most common reactor fuels are composed of either natural or partially enriched uranium. In fewer cases the fresh fuel may be a mix of plutonium and uranium. The contribution of uranium-238 to fission is quite small in most reactors and we ignore it for purposes of illustration. From Eq. (1.28) we observe, however, that if fertile material such as uranium-238 is present in substantial amounts, it will capture neutrons and then undergo radioactive decay to become a fissile material, which may then undergo fission. Thus for a uranium-fueled reactor we must consider fission in both uranium-235 and plutonium-239 (which according to the convention introduced in Section 1.6 we refer to as "25" and "49," respectively). Thus accounting for power production from the two fissile isotopes, we have

$$P''' = \gamma \left[\sigma_f^{25} N^{25}(t) + \sigma_f^{49} N^{49}(t) \right] \phi(t). \qquad (10.26)$$

Note that since the isotopic concentrations change with time as the fissile material is depleted, the flux must also be time dependent, gradually increasing to allow the reactor power to remain constant over long periods of time.

Fissionable Nuclide Concentrations

Rate equations similar to those discussed earlier for the buildup and decay of fission products govern the evolution of the concentrations of fissile and fertile materials. For the uranium isotopes,

$$\frac{d}{dt} N^{25}(t) = -N^{25}(t) \sigma_a^{25} \phi(t), \qquad (10.27)$$

and

$$\frac{d}{dt} N^{28}(t) = -N^{28}(t) \sigma_a^{28} \phi(t). \qquad (10.28)$$

In contrast, the rate equation for plutonium must include a production as well as a destruction term since it is the product of the decay chain given in Eq. (1.28):

$$n + {}^{238}_{92}\text{U} \rightarrow {}^{239}_{92}\text{U} \xrightarrow[23\text{m}]{\beta} {}^{239}_{93}\text{Np} \xrightarrow[2.36\text{d}]{\beta} {}^{239}_{94}\text{Pu}. \tag{10.29}$$

The uranium-238, which constitutes a large fraction of the fuel nuclei in most reactors, captures neutrons and undergoes decay first to neptunium and then to plutonium-239. Since the half-lives of these processes are short compared to the weeks, months, or years over which fuel depletion studies are made, little accuracy is lost by assuming that the plutonium appears instantaneously, following neutron capture by uranium-238. Conversely, although plutonium is radioactive, its half-life is so long that the fraction of plutonium that decays away during the life of a reactor core is negligible. Thus we assume that plutonium is created immediately upon capture of a neutron in uranium-238 and is destroyed only by neutron absorption, which either causes fission or the creation of plutonium-240. In either case the rate equation is

$$\frac{d}{dt}N^{49}(t) = \sigma_\gamma^{28}\phi(t)N^{28}(t) - \sigma_a^{49}\phi(t)N^{49}(t). \tag{10.30}$$

We solve the three rate equations for uranium and plutonium by first integrating Eq. (10.27) directly to obtain

$$N^{25}(t) = N^{25}(0)e^{-\sigma_a^{25}\Phi(t)}. \tag{10.31}$$

where the neutron fluence is defined by

$$\Phi(t) = \int_0^t \phi(t')dt'. \tag{10.32}$$

In the same manner, the solution of Eq. (10.28) is

$$N^{28}(t) = N^{28}(0)e^{-\sigma_a^{28}\Phi(t)}. \tag{10.33}$$

Since the absorption cross section of uranium-238 is quite small compared to the fission cross sections appearing in these equations, however, we may assume that the change in its concentration can be neglected,

$$N^{28}(t) \approx N^{28}(0), \tag{10.34}$$

without substantial loss of accuracy. Applying this approximation reduces Eq. (10.30) to

$$\frac{d}{dt}N^{49}(t) = \sigma_\gamma^{28}\phi(t)N^{28}(0) - \sigma_a^{49}\phi(t)N^{49}(t), \tag{10.35}$$

which is easily solved using the integrating factor technique of Appendix A. Assuming no plutonium is present at the beginning of reactor life, so that $N^{49}(0) = 0$, we obtain

$$N^{49}(t) = \sigma_\gamma^{28} N^{28}(0) \, e^{-\sigma_a^{49}\Phi(t)} \int_0^t e^{\sigma_a^{49}\Phi(t')} \phi(t')dt'. \tag{10.36}$$

Using Eq. (10.32) and the differential transformation $d\Phi = \phi(t')dt'$, the integral may be evaluated to yield

$$N^{49}(t) = \frac{\sigma_\gamma^{28}}{\sigma_a^{49}} N^{28}(0)\left(1 - e^{-\sigma_a^{49}\Phi(t)}\right). \tag{10.37}$$

Figure 10.3 shows the long-term evolution of the slightly enriched fuel typical of that in water-cooled reactors. In addition to the two fissile nuclides treated in our simple model, the figure shows

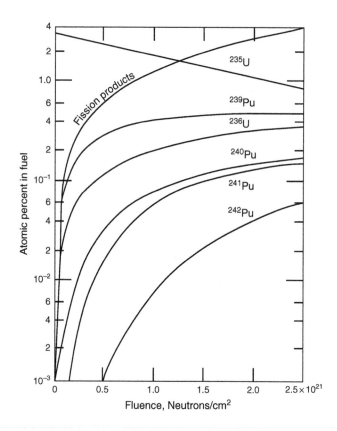

FIGURE 10.3 Change in fuel composition vs neutron fluence (adapted from "FUELCYC, a New Computer Code for Fuel Cycle Analysis," *Nucl. Sci. Eng.*, 11, 386 (1961), by M. Benedict *et al.*, Copyright by the American Nuclear Society, La Grange Park, IL).

the buildup of the higher isotopes of plutonium and uranium that result from successive neutron captures. A quantity that is often used to evaluate fuel performance is the conversion ratio, defined as the ratio of fissile material created to fissile material destroyed. If the ratio is greater than one, it is often referred to as the breeding ratio, for then the reactor is creating more fissile material than it is consuming. In our simplified model, which includes only uranium and plutonium-239, the conversion ratio is

$$CR(t) = \frac{\sigma_\gamma^{28} N^{28}(0)}{\sigma_a^{25} N^{25}(t) + \sigma_a^{49} N^{49}(t)}. \tag{10.38}$$

This equation indicates that increased fuel enrichment results in a decreased value of $CR(0)$, the initial conversion ratio. Thereafter the plutonium isotopes build into the core. As this happens an increasing fraction of the fission comes from plutonium. By the time the right-hand side of Fig. 10.3 is reached more than one-third of the fissile material in the core is plutonim-239.

The effect of the depletion of uranium-235 and the buildup of plutonium on reactivity is illustrated by writing the fuel cross sections in Eq. (10.4) in terms of the nuclide densities:

$$k(t) = \frac{\nu\sigma_f^{25} N^{25}(t) + \nu\sigma_f^{49} N^{49}(t)}{\sigma_a^{25} N^{25}(t) + \sigma_a^{49} N^{49}(t) + \Sigma_\gamma^{28} + \Sigma_\gamma^{fp}(t) + \varsigma(V_m/V_f)\Sigma_a^m} \varepsilon p P_{NL}. \tag{10.39}$$

Generally, the depletion of uranium-235 and the buildup of fission products will overwhelm the buildup of plutonium-239. Thus the value of the multiplication will decrease with time. The net effect of these competing phenomena takes the form similar to that shown in the example labeled "no burnable poison" in Fig. 10.4.

Burnable Poisons

In pressurized water reactors dissolving a soluble neutron absorber in the coolant and varying the concentration with time compensate for much of the excess reactivity. Burnable poisons embedded in the fuel or other core constituents offer an additional means of limiting excess reactivity as well as mitigating localized power peaking.

A burnable poison is an isotope possessing a large neutron capture cross section that is embedded in the fuel or other core constituent to limit excess reactivity early in core life. Boron and gadolinium are often

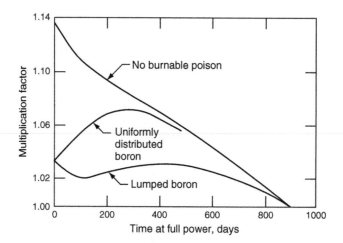

FIGURE 10.4   Effect of burnable poisons on reactor multiplication (adapted from "Kinetics of Solid-Moderator Reactors," by H. B. Steward and M. H. Merrill, *Technology of Nuclear Reactor Safety*, Vol. 1, 1965, T. J. Thompson and J. G. Beckerley, Eds. Countesy of the MIT Press).

used for this purpose. A burnable poison's concentration is limited by a rate equation similar to those above:

$$\frac{d}{dt}N^{BP}(t) = -N^{BP}(t)\sigma_\gamma^{BP}\phi(t), \tag{10.40}$$

which has a solution of

$$N^{BP}(t) = N^{BP}(0)e^{-\sigma_\gamma^{BP}\Phi(t)}. \tag{10.41}$$

Its effect is apparent by adding its capture cross section to the denominator of Eq. (10.39):

$$k(t) = \frac{\nu\sigma_f^{25}N^{25}(t) + \nu\sigma_f^{49}N^{49}(t)}{\sigma_a^{25}N^{25}(t) + \sigma_a^{49}N^{49}(t) + \Sigma_\gamma^{28} + \Sigma_\gamma^{fp}(t) + \varsigma(V_m/V_f)\Sigma_a^m + \sigma_\gamma^{BP}N^{BP}(t)}\varepsilon p P_{NL}. \tag{10.42}$$

Because a burnable poison's capture cross section is large compared to the others appearing in this expression, the exponential in Eq. (10.41) decays rapidly, causing its effect on the multiplication to be concentrated early in core life. The curve marked "uniformly distributed boron" in Fig. 10.4 illustrates this point. Burnable poisons are selected such that the isotope resulting from neutron capture has a small capture cross section, thus avoiding subsequent poisoning of the chain reaction.

Further leveling of the excess reactivity curve can be accomplished by concentrating the burnable poison in lumps instead of distributing it uniformly across the fuel. With such lumping of a strong absorber, spatial self-shielding, similar to that discussed for resonance capture in Chapter 4, takes place. Consequently, neutron capture in the burnable poison is limited early in the fuel life and results in a smoothing of the poison's suppression of excess reactivity over a longer time period. As the curve marked "lumped boron" in Fig. 10.4 illustrates, lumping the poison further minimizes the maximum excess reactivity, and therefore facilitates the design of the control system. Clever design of such lumping can also serve to flatten the spatial distribution of the power density over the life of a core.

10.5 Fission Product and Actinide Inventories

Thus far we have considered the production of fission products and actinides as they affect the reactivity of a power reactor core. The radioactive materials generated during the operation of power reactors also are of concern for the potential health hazards that they represent. We briefly allude to the two basic categories of concern: preventing the radioactive materials from reaching the environment should a catastrophic accident take place while the reactor is operating, and providing for long-term disposal of the radioactive inventory following removal of the fuel from the reactor. The inventories, half-lives, chemical characteristics, and other properties of the isotopes that are most important differ depending on whether reactor safety during core lifetime or long-term disposal of radioactive waste following shutdown is under consideration.

During the core lifetime fission product concentrations build and reach saturation levels according to their yields and characteristic half-lives. To recapitulate earlier discussions, following several half-lives the inventory of a fission product reaches its saturation value. A reactor's inventory of isotopes with half-lives short compared to the operational life of its fuel will be proportional to the reactor's power after it has been operating for some length of time. Conversely, isotopes with half-lives that are long compared to the reactor core life—say, a decade or more—build linearly with time and result in an inventory that is proportion to the amount of energy that the reactor has produced.

Reactor safety is concerned with preventing the accidental release fission products or other radioactive materials to the environment. Fission products with short as well as long half-lives must be safeguarded. Equally important are the quantities in which they are produced, the volatility of the chemical forms that they take, their ability to penetrate the reactor containment and other engineered barriers against their release to the environment, and finally the mechanisms by which they cause biological damage. Radioiodine, the noble gases, strontium, and cesium are among the most important isotopes to be scrutinized in the analysis of hypothesized reactor accidents.

In contrast, only isotopes with long half-lives are of concern for waste disposal, for those with half-lives of a few years or less are easily stored until their radioactivity is exhausted. The more persistent fission products have half-lives of several decades, and these must be isolated for hundreds of years. The actinides—plutonium, neptunium, americium, and others produced by successive capture of neutrons in uranium and its by-products during reactor operation—present the truly long-term challenges in waste disposal. Although

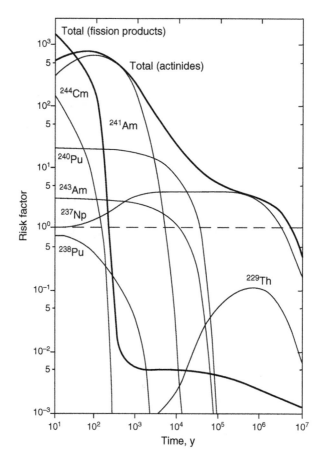

FIGURE 10.5   Time dependence of the risk factor for wastes from spent fuel, stored without reprocessing, normalized to the uranium ore needed to produce the PWR fuel from which it came (adapted from "Formation of Recycling of Minor Actinides in Nuclear Power Stations," by L. Koch, in *Handbook of the Physics of Chemistry of Actinides*, Vol. 4, 1986. Courtesy of Elsevier Science, Amsterdam).

produced in smaller quantities than fission products, actinides have half-lives measured in thousands of years or more. Figure 10.5 shows the radiotoxicity from discharged fuel, relative to that of uranium ore. Note that for time spans of more than a century following reactor shutdown virtually all of the radioactivity results from actinides rather than from fission products. Increasing attention is being directed toward reprocessing reactor fuel in order to recycle it, fissioning more of the plutonium created from uranium-238 capture and increasing greatly the quantity of energy produced relative to the mass of long half-life radioactive waste created.

Bibliography

Benedict, M., et al., "FUELCYC, a New Computer Code for Fuel Cycle Analysis", *Nucl. Sci. Eng.*, **11**, 386 (1961).

Bonalumi, R. A., "In-Core Fuel Management in CANDU-PHR Reactors," *Handbook of Nuclear Reactors Calculations, II*, Yigan Ronen, Ed., CRC Press, Boca Raton, FL, 1986.

Cember, H., "*Introduction to Health Physics*, 3rd ed., McGraw-Hill, NY, 1996.

Cochran, Robert G., and Nicholas Tsoulfanidis, *The Nuclear Fuel Cycle: Analysis and Management*, 2nd ed., American Nuclear Society, La Grange Park, IL, 1999.

Koch, L., "Formation and Recycling of Minor Actinides in Nuclear Power Stations," in *Handbook of the Physics and Chemistry of Actinides*, Vol. 4, A. J. Freeman and C. Keller, Eds., Elsevier Science Publishers, Amsterdam, 1986.

Lamarsh, John R., *Introduction to Nuclear Reactor Theory*, Addison-Wesley, Reading, MA, 1972.

Levine, Samuel H., "In-Core Fuel Management for Four Reactor Types," *Handbook of Nuclear Reactors Calculations, II*, Yigan Ronen, Ed., CRC Press, Boca Raton, FL, 1986.

Lewis, E. E., *Nuclear Power Reactor Safety*, Wiley, NY, 1977.

Salvatores, Max, "Fast Reactor Calculations," *Handbook of Nuclear Reactors Calculations, III*, Yigan Ronen, Ed., CRC Press, Boca Raton, FL, 1986.

Stacey, Weston M., *Nuclear Reactor Physics*, Wiley, NY, 2001.

Steward, H. B., and M. H. Merril, "Kinetics of Solid-Moderator Reactors," *Technology of Nuclear Reactor Safety*, T. J. Thompson and J. G. Beckerley, Eds., Vol. 1, MIT Press, Cambridge, MA, 1966.

Turinsky, Paul J., "Thermal Reactor Calculations," *Handbook of Nuclear Reactors Calculations, III*, Yigan Ronen, Ed., CRC Press, Boca Raton, FL, 1986.

Problems

10.1. Prove that for a reactor operating at a very high flux level, the maximum xenon-135 concentration takes place at approximately 11.3 hours following shutdown.

10.2. Make a logarithmic plot of the effective half-life of xenon-135 over the flux range of $10^{10} \leq \phi \leq 10^{15} \text{n/cm}^2/\text{s}$.

10.3. A thermal reactor fueled with uranium has been operating at constant power for several days. Make a plot of the ratio of concentration of xenon-135 to uranium-235 atoms in the reactor versus its average flux. Determine the maximum value that this ratio can take.

10.4. A pressurized water reactor at full power has an average power density of $\bar{P}''' = 80 \text{ MW/m}^3$ and a peaking factor of $F_q = 2.0$. After the reactor has operated for several days,

a. What is the average xenon concentration?
b. What is the maximum xenon concentration?
c. What is the average samarium concentration?
d. What is the maximum samarium concentration?
(Assume a fission cross section of $\bar{\Sigma}_f = 0.203 \text{ cm}^1$.)

10.5. A reactor is started up and operated at constant power. Solve Eqs. (10.12) and (10.13) and determine the iodine and xenon concentration as a function of time.

10.6. Samarium-157 is produced at a rate of 7.0×10^5 atoms/fission. It then undergoes decay: $^{157}_{62}\text{Sm} \xrightarrow[0.5 \text{ min}]{\beta} {}^{157}_{63}\text{Eu} \xrightarrow[15.2 \text{ hr}]{\beta} {}^{157}_{64}\text{Gd}$. Although the absorption cross section of samarium and europium are negligible, the thermal absorption cross section of gadolinium is 240,000 b. Suppose that a reactor operates at a power density of 100 MW/m^3 and a flux level of $8.0 \times 10^{12} \text{ n/cm}^2/\text{s}$.

a. Solve the decay equations for $G(t)$, the atom density of gadolinium, at a time t following reactor startup.
b. Evaluate $G(\infty)$.
c. If the reactor has been operated for several weeks and then is shut down, what is the concentration of gadolinium, after the reactor has been shut down for several weeks?
(Assume that the energy produced per fission is $3.1 \times 10^{-11} \text{ W s}$)

10.7. Verify Eqs. (10.18) and (10.19).

10.8. A reactor has operated for several weeks at constant power, reaching the equilibrium concentrations of iodine and xenon I_o and X_o given by Eqs. (10.14) and (10.15). At $t = 0$ the power is cut back, dropping the flux level from ϕ to $\tilde{\phi}$. Solve Eqs. (10.12) and (10.13) and show that the iodine and xenon concentrations following the power reduction are

$$I(t) = \frac{\gamma_I}{\lambda_I} \bar{\Sigma}_f \left[\tilde{\phi} + (\phi - \tilde{\phi})e^{-\lambda_I t} \right]$$

and

$$X(t) = \frac{(\gamma_I + \gamma_X)}{\lambda_X + \sigma_{aX}\phi} \bar{\Sigma}_f \phi e^{-(\lambda_X + \sigma_{aX}\tilde{\phi})t} + \frac{(\gamma_I + \gamma_X)}{\lambda_X + \sigma_{aX}\tilde{\phi}} \bar{\Sigma}_f \tilde{\phi} \left[1 - e^{-(\lambda_X + \sigma_{aX}\tilde{\phi})t} \right]$$
$$+ \frac{\gamma_I}{\lambda_X - \lambda_1 + \sigma_{aX}\tilde{\phi}} \bar{\Sigma}_f (\phi - \tilde{\phi}) \left[e^{-\lambda_I t} - e^{-(\lambda_X + \sigma_{aX}\tilde{\phi})t} \right].$$

10.9. Under load following conditions, a reactor operates each day at full power for 12 hours, followed by a shutdown of 12 hours. Calculate the iodine concentration, $I(t)$, over a 24 hour time span. Use the periodic boundary condition $I(24 \ hr) = I(0)$.

10.10. Under load following conditions, a reactor operates each day at full power for 12 hours, followed by a shutdown of 12 hours. Calculate the promethium concentration, $P(t)$, over a 24 hour time span. Use the periodic boundary condition $P(24 \ hr) = P(0)$.

10.11. Taking into account neutron capture in plutonium-239 and plutonium-240,

   a. Write a rate equation for the concentration of plutonium-240.
   b. Solve the equation from part a using Eq. (10.37) for the concentration of plutonium-239.

10.12. Neptunium-238 has a thermal absorption cross section of 33 b, which we have neglected in deriving Eq. (10.30). In a reactor operating at a flux level of $\phi = 5 \times 10^{14} \, \text{n/cm}^2/\text{s}$, what fraction of the neptunium will capture a neutron instead of decaying to plutonium-239?

10.13. Consider uranium fuel in a thermal reactor with an initial enrichment of 4%.

   a. What is the conversion ratio (CR) at the beginning of core life?
   b. After half of the uranium-235 has been burned, what is the conversion ratio?
   c. After half of the uranium-235 has been burned, what fraction of the power is being produced from plutonium-239?

   Hint: Make use of the approximations in Eqs. (10.31) and (10.37).

10.14. Thorium-232 is a fertile material that may be transmuted to fissile uranium-233 through the following reaction: $^{232}_{90}\text{Th} \xrightarrow{n} {}^{233}_{90}\text{Th} \xrightarrow[22\,\text{min}]{\beta} {}^{233}_{91}\text{Pa} \xrightarrow[27.4\,\text{days}]{\beta} {}^{233}_{92}\text{U}$, where the half-lives are indicated. Assume that a fresh core is put into operation containing only thorium-232 and uranium-235. Thereafter neutron capture in thorium takes place at a constant rate of $\Sigma_a^{th}\bar{\phi}$.

a. Assuming that the half-life of $^{233}_{90}\text{Th}$ can be ignored, write down and solve the differential equation for the concentration of $^{233}_{91}\text{Pa}$.

b. Write down and solve the differential equation for the concentration of $^{233}_{92}\text{U}$.

(Assume $N^{02}(t) = N^{02}(0)$ and $\phi(t) = \phi(0)$.)

Appendices

APPENDIX A

Useful Mathematical Relationships

Derivatives and Integrals

$$\frac{d}{dx}x^n = nx^{n-1}$$

$$\int x^n dx = \begin{cases} \frac{1}{n+1}x^{n+1} + C, & n \neq -1 \\ \ln(x) + C, & n = -1 \end{cases}$$

$$\frac{d}{dx}\ln(x) = x^{-1}$$

$$\int \ln(x)dx = x\ln(x) - x + C$$

$$\frac{d}{dx}e^{ax} = ae^{ax}$$

$$\int e^{ax}dx = \frac{1}{a}e^{ax} + C$$

$$\frac{d}{dx}\sin(ax) = a\cos(ax)$$

$$\int \sin(ax)dx = -\frac{1}{a}\cos(ax) + C$$

$$\frac{d}{dx}\cos(ax) = -a\sin(ax)$$

$$\int \cos(ax)dx = \frac{1}{a}\sin(ax) + C$$

$$\frac{d}{dx}\sinh(ax) = a\cosh(ax)$$

$$\int \sinh(ax)dx = \frac{1}{a}\cosh(ax) + C$$

$$\frac{d}{dx}\cosh(ax) = a\sinh(ax)$$

$$\int \cosh(ax)dx = \frac{1}{a}\sinh(ax) + C$$

Definite Integrals

$$\int_0^\infty x^n e^{-x}dx = n! \qquad \int_0^\infty e^{-x^2}dx = \frac{1}{2}\sqrt{\pi}$$

$$\int_0^\infty xe^{-x^2}dx = \frac{1}{2} \qquad \int_0^\infty x^2 e^{-x^2}dx = \frac{1}{4}\sqrt{\pi}$$

$$\int_0^\infty x^{2n}e^{-x^2}dx = \frac{1}{2^{n+1}}[1 \cdot 3 \cdot 5 \cdots (2n-1)]\sqrt{\pi}, \quad n = 1, 2, 3, \cdots$$

Hyperbolic Functions

$$\sinh(ax) = \frac{1}{2}(e^{ax} - e^{-ax}) \qquad \cosh(ax) = \frac{1}{2}(e^{ax} + e^{-ax})$$

$$\tanh(ax) = \sinh(ax)/\cosh(ax) \quad \coth(ax) = \cosh(ax)/\sinh(ax)$$

Expansions

$$\sin x = x - \frac{x^3}{3!} + \frac{x^5}{5!} - \frac{x^7}{7!} + \cdots \qquad |x| < \infty$$

$$\cos x = 1 - \frac{x^2}{2!} + \frac{x^4}{4!} - \frac{x^6}{6!} + \cdots \qquad |x| < \infty$$

$$\tan x = x + \frac{x^3}{3} + \frac{2}{15}x^5 + \cdots \qquad |x| < \pi/2$$

$$\cot x = \frac{1}{x} - \frac{x}{3} - \frac{x^3}{45} - \cdots \qquad |x| < \pi$$

$$e^x = 1 + \frac{x}{1!} + \frac{x^2}{2!} + \frac{x^3}{3!} + \frac{x^4}{4!} + \cdots \qquad |x| < \infty$$

$$\sinh x = x + \frac{x^3}{3!} + \frac{x^5}{5!} + \frac{x^7}{7!} + \cdots \qquad |x| < \infty$$

$$\cosh x = 1 + \frac{x^2}{2!} + \frac{x^4}{4!} + \frac{x^6}{6!} + \cdots \qquad |x| < \infty$$

$$\tanh x = x - \frac{x^3}{3} + \frac{2}{15}x^5 - \cdots \qquad |x| < \pi/2$$

$$\coth x = \frac{1}{x} + \frac{x}{3} - \frac{x^3}{45} - \cdots \qquad |x| < \pi$$

Integration by Parts

$$\int_a^b f(x)\frac{d}{dx}g(x)dx = f(b)g(b) - f(a)g(a) - \int_a^b g(x)\frac{d}{dx}f(x)dx$$

Derivative of an Integral

$$\frac{d}{dc}\int_p^q f(x,c)dx = \int_p^q \frac{\partial}{\partial c}f(x,c)dx + f(q,c)\frac{dq}{dc} - f(p,c)\frac{dp}{dc}$$

First-Order Differential Equations

$$\frac{d}{dt}y(t) + \alpha(t)y(t) = S(t),$$

where $\alpha(t)$ and $S(t)$ are known. Note that

$$\frac{d}{dt}\left[y(t)e^{\int_0^t \alpha(t')dt'}\right] = \left[\frac{d}{dt}y(t) + \alpha(t)\right]e^{\int_0^t \alpha(t')dt'}.$$

Thus, multiply both sides by the integrating factor $\exp[\int_0^t \alpha(t')dt']$ to obtain

$$\frac{d}{dt}\left[y(t)e^{\int_0^t \alpha(t')dt'}\right] = S(t)e^{\int_0^t \alpha(t')dt'}.$$

Integrating between 0 and t then yields

$$y(t)e^{\int_0^t \alpha(t')dt'} - y(0) = \int_0^t S(t')e^{\int_0^{t'} \alpha(t'')dt''} dt'$$

and solving for $y(t)$ gives

$$y(t) = y(0)e^{-\int_0^t \alpha(t')dt'} + \int_0^t S(t')e^{-\int_{t'}^t \alpha(t'')dt''} dt'.$$

If α is a constant the solution reduces to

$$y(t) = y(0)e^{-\alpha t} + \int_0^t S(t')e^{-\alpha(t-t')}dt'.$$

If S is also a constant then

$$y(t) = y(0)e^{-\alpha t} + \frac{S}{\alpha}(1 - e^{-\alpha t}).$$

Second-Order Differential Equations

$$\frac{d^2}{dx^2} y(x) + a^2 y(x) = S(x)$$

and

$$\frac{d^2}{dx^2} y(x) - b^2 y(x) = S(x),$$

where a and b are constants, and $S(x)$ is known. The solutions are a superposition of a general and a particular solution:

$$y(t) = y_g(t) + y_p(t). \qquad (A.1)$$

The general solutions are those for which the right-hand sides of the equations are set equal to zero. By substitution, the following may be shown to satisfy the equations:

$$y_g(t) = C_1 \sin(ax) + C_2 \cos(ax)$$

and

$$y_g(t) = C_1 \exp(bx) + C_2 \exp(-bx),$$

respectively, where C_1 and C_2 are arbitrary constants. This solution may also be expressed in terms of the hyperbolic functions:

$$y_g(t) = C_1' \sinh(bx) + C_2' \cosh(bx).$$

If S is a constant then the particular solutions are simply $y_p(t) = S/a^2$ or $y_p(t) = -S/b^2$, respectively. After determining the particular solution, two boundary conditions must be applied to Eq. (A.1) to determine C_1 and C_2.

∇^2 and dV in Various Coordinate Systems

Cartesian, one-dimensional: $\quad \nabla^2 = \dfrac{\partial^2}{\partial x^2}$ $\qquad\qquad\qquad dV = \text{area} \times dx$

Cylindrical, one-dimensional: $\quad \nabla^2 = \dfrac{1}{r}\dfrac{\partial}{\partial r} r \dfrac{\partial}{\partial r}$ $\qquad\qquad dV = \text{height} \times 2\pi r dr$

Spherical, one-dimensional: $\quad \nabla^2 = \dfrac{1}{r^2}\dfrac{\partial}{\partial r} r^2 \dfrac{\partial}{\partial r}$ $\qquad\qquad dV = 4\pi r^2 dr$

Cylindrical, two-dimensional: $\quad \nabla^2 = \dfrac{1}{r}\dfrac{\partial}{\partial r} r \dfrac{\partial}{\partial r} + \dfrac{\partial^2}{\partial z^2}$ $\qquad dV = 2\pi r dr dz$

Cartesian, three-dimensional: $\quad \nabla^2 = \dfrac{\partial^2}{\partial x^2} + \dfrac{\partial^2}{\partial y^2} + \dfrac{\partial^2}{\partial x^2}$ $\quad dV = dx dy dz$

APPENDIX B

Bessel's Equation and Functions

The solution of Bessel's equation is frequently required in dealing with neutron distributions in cylindrical geometry. The Bessel equation of order n is

$$\frac{d^2\Phi}{dr^2} + \frac{1}{r}\frac{d\Phi}{dr} + \left(\alpha^2 - \frac{n^2}{r^2}\right)\Phi = 0.$$

The solution to this equation is

$$\Phi(r) = C_1 J_n(\alpha r) + C_2 Y_n(\alpha r),$$

where C_1 and C_2 are arbitrary constants and $J_n(\alpha r)$ and $Y_n(\alpha r)$ are the ordinary Bessel function of the first and second kind, respectively.

The second form of the Bessel equation results when the sign of the parameter α^2 is changed:

$$\frac{d^2\Phi}{dr^2} + \frac{1}{r}\frac{d\Phi}{dr} - \left(\alpha^2 + \frac{n^2}{r^2}\right)\Phi = 0.$$

The solution to this equation is

$$\Phi(r) = C_1' I_n(\alpha r) + C_2' K_n(\alpha r),$$

where the arbitrary constants are now C_1' and C_2', and $I_n(\alpha r)$ and $K_n(\alpha r)$ are known as the modified Bessel function of the first and second kinds, respectively.

In reactor physics calculations Bessel equations of order zero ($n = 0$) most frequently appear. The ordinary and modified Bessel functions of order zero are plotted in Fig. B.1. To apply boundary conditions and to determine integral parameters it is often necessary to differentiate or integrate the zero-order Bessel functions. The required relationships are

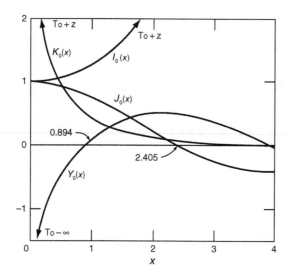

FIGURE B.1   Ordinary and modified Bessel functions of order zero.

$$\frac{d}{dr}J_0(\alpha r) = -\alpha J_1(\alpha r) \qquad \frac{d}{dr}Y_0(\alpha r) = -\alpha Y_1(\alpha r),$$

$$\frac{d}{dr}I_0(\alpha r) = \alpha I_1(\alpha r) \qquad \frac{d}{dr}K_0(\alpha r) = -\alpha K_1(\alpha r),$$

$$\int J_0(\alpha r)r dr = \frac{r}{\alpha}J_1(\alpha r) + C \qquad \int Y_0(\alpha r)r dr = \frac{r}{\alpha}Y_1(\alpha r) + C,$$

$$\int I_0(\alpha r)r dr = \frac{r}{\alpha}I_1(\alpha r) + C \qquad \int K_0(\alpha r)r dr = -\frac{r}{\alpha}K_1(\alpha r) + C.$$

Values of ordinary and modified Bessel functions of order zero and one are tabulated in Table B.1. The values of Y_n and K_n become infinite as the argument goes to zero:

$$Y_0(\alpha r) \xrightarrow[\alpha r \to 0]{} \frac{2}{\pi}[\ln(\alpha r) - 0.11593] \quad Y_1(\alpha r) \xrightarrow[\alpha r \to 0]{} \frac{2}{\pi \alpha r},$$

$$K_0(\alpha r) \xrightarrow[\alpha r \to 0]{} -[\ln(\alpha r) - 0.11593] \quad K_1(\alpha r) \xrightarrow[\alpha r \to 0]{} \frac{1}{\alpha r}.$$

The values of I_n become infinite as the argument goes to infinity:

$$I_n(\alpha r) \xrightarrow[\alpha r \to \infty]{} \frac{1}{\sqrt{2\pi \alpha r}}\exp(\alpha r).$$

TABLE B.1
Bessel Functions of Order Zero and One

X	$I_0(x)$	$I_1(x)$	$Y_0(x)$	$Y_1(x)$	$I_0(x)$	$I_1(x)$	$K_0(x)$	$K_1(x)$
0	1.0000	0.0000	$-\infty$	$-\infty$	1.000	0.0000	∞	∞
0.05	0.9994	0.0250	−1.979	−12.79	1.001	0.0250	3.114	19.91
0.10	0.9975	0.0499	−1.534	−6.459	1.003	0.0501	2.427	9.854
0.15	0.9944	0.0748	−1.271	−4.364	1.006	0.0752	2.030	6.477
0.20	0.9900	0.0995	−1.081	−3.324	1.010	0.1005	1.753	4.776
0.25	0.9844	0.1240	−0.9316	−2.704	1.016	0.1260	1.542	3.747
0.30	0.9776	0.1483	−0.8073	−2.293	1.023	0.1517	1.372	3.056
0.35	0.9696	0.1723	−0.7003	−2.000	1.031	0.1777	1.233	2.559
0.40	0.9604	0.1960	−0.6060	−1.781	1.040	0.2040	1.115	2.184
0.45	0.9500	0.2194	−0.5214	−1.610	1.051	0.2307	1.013	1.892
0.50	0.9385	0.2423	−0.4445	−1.471	1.063	0.2579	0.9244	1.656
0.55	0.9258	0.2647	−0.3739	−1.357	1.077	0.2855	0.8466	1.464
0.60	0.9120	0.2867	−0.3085	−1.260	1.092	0.3137	0.7775	1.303
0.65	0.8971	0.3081	−0.2476	−1.177	1.108	0.3425	0.7159	1.167
0.70	0.8812	0.3290	−0.1907	−1.103	1.126	0.3719	0.6605	1.050
0.75	0.8642	0.3492	−0.1372	−1.038	1.146	0.4020	0.6106	0.9496
0.80	0.8463	0.3688	−0.0868	−0.9781	1.167	0.4329	0.5653	0.8618
0.85	0.8274	0.3878	−0.0393	−0.9236	1.189	0.4646	0.5242	0.7847
0.90	0.8075	0.4059	−0.0056	−0.8731	1.213	0.4971	0.4867	0.7165
0.95	0.7868	0.4234	0.0481	−0.8258	1.239	0.5306	0.4524	0.6560
1.0	0.7652	0.4401	0.0883	−0.7812	1.266	0.5652	0.4210	0.6019
1.1	0.6957	0.4850	0.1622	−0.6981	1.326	0.6375	0.3656	0.5098
1.2	0.6711	0.4983	0.2281	−0.6211	1.394	0.7147	0.3185	0.4346
1.3	0.5937	0.5325	0.2865	−0.5485	1.469	0.7973	0.2782	0.3725
1.4	0.5669	0.5419	0.3379	−0.4791	1.553	0.8861	0.2437	0.3208

(continued)

Table B.1
(continued)

X	$I_0(x)$	$I_1(x)$	$Y_0(x)$	$Y_1(x)$	$I_0(x)$	$I_1(x)$	$K_0(x)$	$K_1(x)$
1.5	0.4838	0.5644	0.3824	-0.4123	1.647	0.9817	0.2138	0.2774
1.6	0.4554	0.5699	0.4204	-0.3476	1.750	1.085	0.1880	0.2406
1.7	0.3690	0.5802	0.4520	-0.2847	1.864	1.196	0.1655	0.2094
1.8	0.3400	0.5815	0.4774	-0.2237	1.990	1.317	0.1459	0.1826
1.9	0.2528	0.5794	0.4968	-0.1644	2.128	1.448	0.1288	0.1597
2.0	0.2239	0.5767	0.5104	-0.1070	2.280	1.591	0.1139	0.1399
2.1	0.1383	0.5626	0.5183	-0.0517	2.446	1.745	0.1008	0.1227
2.2	0.1104	0.5560	0.5208	-0.0015	2.629	1.914	0.0893	0.1079
2.3	0.0288	0.5305	0.5181	0.0523	2.830	2.098	0.0791	0.0950
2.4	0.0025	0.5202	0.5104	0.1005	3.049	2.298	0.0702	0.0837
2.5	0.0729	0.4843	0.4981	0.1459	3.290	2.517	0.0623	0.0739
2.6	-0.0968	0.4708	0.4813	0.1884	3.553	2.755	0.0554	0.0653
2.7	-0.1641	0.4260	0.4605	0.2276	3.842	3.016	0.0493	0.0577
2.8	-0.1850	0.4097	0.4359	0.2635	4.157	3.301	0.0438	0.0511
2.9	-0.2426	0.3575	0.4079	0.2959	4.503	3.613	0.0390	0.0453
3.0	-0.2601	0.3391	0.3769	0.3247	4.881	3.953	0.0347	0.0402
3.2	-0.3202	0.2613	0.3071	0.3707	5.747	4.734	0.0276	0.0316
3.4	-0.3643	0.1792	0.2296	0.4010	6.785	5.670	0.0220	0.0250
3.6	-0.3918	0.0955	0.1477	0.4154	8.028	6.793	0.0175	0.0198
3.8	-0.4026	0.0128	0.0645	0.4141	9.517	8.140	0.0140	0.0157
4.0	-0.3971	-0.0660	-0.0169	0.3979	11.302	9.759	0.0112	0.0125

APPENDIX C

Derivation of Neutron Diffusion Properties

In considering neutron distributions we have examined the effects of neutron energy, and of time dependence. Through the use of the diffusion equation we have been able to examine the spatial distribution of neutrons without bringing into account explicitly the direction of neutron travel. However, in order to derive an expression for the difusion coefficient as well as for the partial currents utilized in conjunction with diffusion theory, we must examine the distribution of neutrons in angle. This requires that we make use of the neutron transport equation. We limit consideration here to the time-independent case; we also eliminate the energy variable by assuming that energy averaging of cross sections has already been accomplished. Thus we can consider all neutrons to travel at an average speed, v. We first derive the monoenergetic time-independent transport equation. We then reduce it to plane geometry and derive the desired quantities.

The Transport Equation

Consider neutrons located in an incremental volume dV located about a point at $\vec{r}$ traveling in cone of directions $d\Omega$ about the direction $\hat{\Omega}$, as indicated in Fig. C.1. Here $dV = dx\,dy\,dz$ and in a polar coordinate system for angle $d\Omega = \sin(\theta)d\theta d\omega$. Hence,

$$\int d\Omega = \int_0^\pi \sin(\theta)d\theta \int_0^{2\pi} d\omega = 4\pi. \qquad (C.1)$$

Suppose $N(\vec{r}, \hat{\Omega})$ is the number of neutrons at $\vec{r}$ traveling in the direction $\hat{\Omega}$. To obtain the flux—also called scalar flux—found in the diffusion equation we integrate over all possible angles and multiply by v, the neutron speed:

$$\phi(\vec{r}) = v \int N(\vec{r}, \hat{\Omega})\,d\Omega. \qquad (C.2)$$

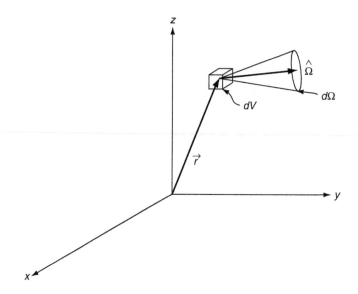

FIGURE C.1 Streaming of neutrons through an incremental volume in the direction $\hat{\Omega}$.

The current is a vector, given by

$$\vec{J}(\vec{r}) = v \int \hat{\Omega} N(\vec{r}, \hat{\Omega})\, d\Omega. \tag{C.3}$$

Thus if $\hat{n}$ is the normal to a surface, then

$$\hat{n} \cdot \vec{J}(\vec{r}) = v \int \hat{n} \cdot \hat{\Omega} N(\vec{r}, \hat{\Omega})\, d\Omega \tag{C.4}$$

is the net number of neutrons crossing the surface in the positive direction/s/cm^2.

    To derive the transport equation we employ the cylindrical volume element shown in Fig. C.2 with its base centered at $\vec{r}$ and its axis parallel to $\hat{\Omega}$, the direction of neutron travel. As shown, it has a height Δu, cross-sectional area ΔA, and hence a volume $\Delta V = \Delta u \Delta A$. The balance equation for neutrons traveling in direction $\hat{\Omega}$ is then

 # neutrons leaving from the right − # neuton sentering from the left

= −# collisions in ΔV + # neutrons emitted in direction $\hat{\Omega}$ in ΔV.

$$\tag{C.5}$$

Note that any collision removes a neutron from direction $\hat{\Omega}$ either by absorption or by scattering it to a different direction $\hat{\Omega}'$. Neutrons may be emitted into direction $\hat{\Omega}$ either as a result of scattering or directly from a source. We may write Eq. (C.5) in terms of $N(\vec{r}, \hat{\Omega})$ as

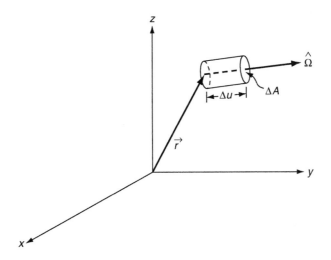

FIGURE C.2 Neutrons in volume dV traveling the cone $d\Omega$ about the direction $\hat{\Omega}$.

$$vN(\vec{r} + \Delta u, \hat{\Omega})\Delta A - vN(\vec{r}, \hat{\Omega})\Delta A$$
$$= -\Sigma_t(\vec{r})vN(\vec{r}, \hat{\Omega})\Delta u\Delta A + Q(r)\Delta u\Delta A. \qquad \text{(C.6)}$$

Neutron emission from isotropic scattering and sources may be written as

$$Q(\vec{r}) = \frac{1}{4\pi}\Sigma_s(\vec{r})\phi(\vec{r}) + \frac{1}{4\pi}s(\vec{r}). \qquad \text{(C.7)}$$

Dividing Eq. (C.6) by $\Delta u\Delta A$ and noting that

$$\frac{vN(\vec{r} + \Delta u, \hat{\Omega}) - vN(\vec{r}, \hat{\Omega})}{\Delta u} \rightarrow \frac{d}{du}vN(\vec{r}, \hat{\Omega}), \qquad \text{(C.8)}$$

we obtain

$$\frac{d}{du}vN(\vec{r}, \hat{\Omega}) = -\Sigma_t(\vec{r})vN(\vec{r}, \hat{\Omega}) + \frac{1}{4\pi}\Sigma_s(\vec{r})\phi(\vec{r}) + \frac{1}{4\pi}s(\vec{r}). \quad \text{(C.9)}$$

We proceed by writing the derivative on the left as

$$\frac{d}{du} = \frac{\partial}{\partial x}\frac{dx}{du} + \frac{\partial}{\partial y}\frac{dy}{du} + \frac{\partial}{\partial z}\frac{dz}{du}$$
$$= \hat{\Omega} \cdot \hat{i}\frac{\partial}{\partial x} + \hat{\Omega} \cdot \hat{j}\frac{\partial}{\partial y} + \hat{\Omega} \cdot \hat{k}\frac{\partial}{\partial z} = \hat{\Omega} \cdot \vec{\nabla}. \qquad \text{(C.10)}$$

Hence, Eq. (C.9) becomes

$$\hat{\Omega} \cdot \vec{\nabla} v N(\vec{r}, \hat{\Omega}) + \Sigma_t(\vec{r}) v N(\vec{r}, \hat{\Omega}) = \frac{1}{4\pi} \Sigma_s(\vec{r}) \phi(\vec{r}) + \frac{1}{4\pi} s(\vec{r}). \quad \text{(C.11)}$$

We next impose the symmetry of plane geometry on this equation. The spatial variation depends only on x and the angular variation only on $\mu = \hat{\Omega} \cdot \hat{i} = \cos(\theta)$, the direction cosine of the x axis. We then have $N(\vec{r}, \hat{\Omega}) \to N(x, \mu)$ and

$$\hat{\Omega} \cdot \vec{\nabla} \to \mu \frac{\partial}{\partial x}. \quad \text{(C.12)}$$

With these substitutions we can integrate Eq. (C.11) over the azimuthal angle, ω, to obtain

$$\mu \frac{\partial}{\partial x} \Psi(x, \mu) + \Sigma_t(x) \Psi(x, \mu) = \frac{1}{2} \Sigma_s(x) \phi(x) + \frac{1}{2} s(x), \quad \text{(C.13)}$$

where

$$\Psi(x, \mu) = \int_0^{2\pi} v N(x, \mu, \omega) d\omega \quad \text{(C.14)}$$

and

$$\phi(x) = \int_{-1}^{1} \Psi(x, \mu) d\mu. \quad \text{(C.15)}$$

The Diffusion Approximation

To obtain the diffusion coefficient we assume that the angular variation in Ψ is linear in μ. It is then easily shown that

$$\Psi(x, \mu) = \frac{1}{2} \phi(x) + \frac{3}{2} \mu J(x), \quad \text{(C.16)}$$

because, in plane geometry, Eq. (C.4) reduces to

$$\hat{i} \cdot \vec{J}(\vec{r}) \equiv J(x) = \int_{-1}^{1} \mu \Psi(x, \mu) d\mu. \quad \text{(C.17)}$$

We cannot solve Eq. (C.13) exactly with the approximation of Eq. (C.16). However, we can require that the equation be satisfied in the following weighted residual sense. We substitute Eq. (C.16) into (C.13) after weighting it by $w(\mu)$ and integrating over μ between -1 and $+1$. The result is

$$\int_{-1}^{1} d\mu w(\mu) \left[\frac{1}{2}\mu \frac{d}{dx}\phi(x) + \frac{3}{2}\mu^2 \frac{d}{dx}J(x) + \frac{1}{2}\Sigma_t(x)\phi(x) + \frac{3}{2}\mu\Sigma_t(x)J(x) \right]$$
$$= \int_{-1}^{1} d\mu w(\mu)\frac{1}{2}[\Sigma_s(x)\phi(x) + s(x)].$$

(C.18)

We first take $w(\mu) = 1$, causing the terms that are odd in μ to vanish and yielding the neutron balance equation:

$$\frac{d}{dx}J(x) + \Sigma_t(x)\phi(x) = \Sigma_s(x)\phi(x) + s(x). \qquad (C.19)$$

Conversely, taking $w(\mu) = \mu$ causes the even terms in μ to vanish and yields

$$\frac{1}{3}\frac{d}{dx}\phi(x) + \Sigma_t(x)J(x) = 0. \qquad (C.20)$$

This is just the one-dimensional form of Fick's law,

$$J(x) = -D(x)\frac{d}{dx}\phi(x). \qquad (C.21)$$

Comparing Eqs. (C.20) and (C.21), we see that the diffusion coefficient is defined by

$$D(x) = 1/3\Sigma_t(x). \qquad (C.22)$$

We have assumed isotropic scattering in this derivation. For anisotropic scattering the term on the right of Eq. (C.13) and subsequent equations includes the anisotropic contribution. The diffusion coefficient becomes

$$D(x) = 1/3\Sigma_{tr}(x), \qquad (C.23)$$

where the transport cross section is given by

$$\Sigma_{tr}(x) = \Sigma_t(x) - \bar{\mu}\Sigma_s(x), \qquad (C.24)$$

where $\bar{\mu}$ is the averaged value of the cosine of the scattering angle.

Partial Currents

We define the partial currents $J^{\pm}(x)$, respectively, as the number of neutrons traveling to the right and to the left through a plane perpendicular to the x axis. From $J(x)$ defined in Eq. (C.17) we see that $\mu > 0$ for neutrons traveling to the right and hence

$$J^{+}(x) = \int_{0}^{1} \mu \Psi(x, \mu) d\mu, \qquad (C.25)$$

whereas $\mu < 0$ for neutrons traveling to the left, and thus

$$J^{-}(x) = \int_{-1}^{0} |\mu| \Psi(x, \mu) d\mu. \qquad (C.26)$$

The net number of neutrons passing to the right is then

$$J(x) = J^{+}(x) - J^{-}(x). \qquad (C.27)$$

We employ the diffusion approximation to derive the partial currents in terms of the flux and net current. Inserting Eq. (C.16) into Eqs. (C.25) and (C.26) and performing the integrals over μ:

$$J^{\pm}(x, \mu) = \frac{1}{4}\phi(x) \pm \frac{1}{2}J(x), \qquad (C.28)$$

or using Eq. (C.21) to eliminate the current,

$$J^{\pm}(x, \mu) = \frac{1}{4}\phi(x) \mp \frac{1}{2}D(x)\frac{d}{dx}\phi(x). \qquad (C.29)$$

APPENDIX D

Fuel Element Heat Transfer

The diameters of cylindrical fuel elements are very small relative to their length; thus heat transfer in the axial direction can be neglected and treated as one-dimensional in the radial directions. In cylindrical geometry the heat transfer equation in the fuel region is

$$k_f \frac{1}{\varsigma} \frac{d}{d\varsigma} \varsigma \frac{d}{d\varsigma} T(\varsigma) + q''' = 0, \tag{D.1}$$

where we assume that the thermal conductivity k_f is a constant. We also assume the heat per unit volume q''' to be radially uniform over the fuel region. Thus we can express it in terms of the linear heat rate q' as

$$q''' = \frac{q'}{\pi a^2}, \tag{D.2}$$

where a is the radius of the fuel regions indicated in Fig. D.1. Equation (D.1) can be integrated twice to yield

$$T(\varsigma) = -\frac{q' \varsigma^2}{4\pi a^2 k_f} + C_1 \ln \varsigma + C_2 \tag{D.3}$$

where C_1 and C_2 are the constants of integration. Clearly, $C_1 = 0$ because otherwise the temperature would be infinite at the centerline; C_2 is determined by requiring that the temperature take on a value of $T(a^-)$ on the fuel side of the fuel–cladding interface. Performing the necessary algebra yields

$$T(\varsigma) = T(a^-) + \frac{q'}{4\pi k_f} \left[1 - \left(\frac{\varsigma}{a} \right)^2 \right], \qquad 0 \le \varsigma \le a. \tag{D.4}$$

We require both the average and the maximum fuel temperature. The radial average over the cylindrical volume is calculated from

$$\bar{T}_f = \frac{2}{a^2} \int_0^a T(\varsigma)\varsigma \, d\varsigma. \tag{D.5}$$

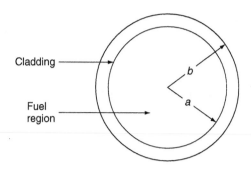

FIGURE D.1   Cylindrical fuel element.

Thus, utilizing Eq. (D.4) we have

$$\bar{T}_f = T(a^-) + \frac{q'}{8\pi k_f}. \tag{D.6}$$

The highest temperature occurs at the centerline where $\varsigma = 0$. Hence, from Eq. (D.4),

$$T_{cl} = T(0) = T(a^-) + \frac{q'}{4\pi k_f}. \tag{D.7}$$

Typically there is a significant thermal resistance across the fuel–cladding interface. Such resistance may be expressed as a gap heat transfer coefficient h_g, defined in terms of the heat flux per unit area q_g'' and the temperature drop across the gap:

$$q_g'' = h_g[T(a^-) - T(a^+)]. \tag{D.8}$$

Because the total heat flux across the gap must equal the linear heat rate,

$$q' = 2\pi a q_g'', \tag{D.9}$$

we may write

$$T(a^-) - T(a^+) = \frac{q'}{2\pi a h_g}. \tag{D.10}$$

We obtain the temperature distribution across the cladding from the source-free heat conduction equation in cylindrical geometry:

$$k_c \frac{1}{\varsigma} \frac{d}{d\varsigma} \varsigma \frac{d}{d\varsigma} T(\varsigma) = 0, \tag{D.11}$$

where k_c is the cladding thermal conductivity. Integrating twice, we obtain

$$T(\varsigma) = C_1 \ln \varsigma + C_2, \qquad a \leq \varsigma \leq b. \tag{D.12}$$

We determine the integration constants C_1 and C_2 from the heat flux at $\varsigma = a^+$ as well as the temperature at $\varsigma = a^+$. From Eq. (D.9) and the Fourier law of heat conduction, the heat flux is

$$\frac{q'}{2\pi a} = -k_c \frac{d}{d\varsigma} T(\varsigma) \Big|_{\varsigma = a^+}. \tag{D.13}$$

Evaluating the constants in Eq. (D.12) then yields

$$T(\varsigma) = T(a^+) - \frac{q'}{2\pi k_c} \ln \left(\frac{\varsigma}{a} \right), \qquad a \leq \varsigma \leq b. \tag{D.14}$$

If the cladding is thin relative to the fuel radius $\tau \equiv b - a \ll a$, we may approximate the logarithm as $\ln(b/a) \simeq \tau/a$, allowing us to evaluate Eq. (D.14) at $\varsigma = b$ and write

$$T(a^+) - T(b) = \frac{\tau}{2\pi a k_c} q'. \tag{D.15}$$

The temperature drop between the cladding surface and the coolant temperature—averaged over the cross-sectional area of the channel—may be determined from the heat transfer coefficient defined by

$$q_c'' = h[T(b) - T_c]. \tag{D.16}$$

Because the total heat flux into the coolant must equal the linear heat rate, $q' = 2\pi b q_c''$, we may write

$$T(b) - T_c = \frac{q'}{2\pi b h}. \tag{D.17}$$

To find R'_{fe}, the fuel element thermal resistance per unit length, we add the temperature drops from Eqs. (D.6), (D.10), (D.15), and (D.17) to obtain

$$T_f - T_c = R'_f q', \tag{D.18}$$

where

$$R'_{fe} = \frac{1}{8\pi k_f} + \frac{1}{2\pi a h_g} + \frac{\tau}{2\pi a k_c} + \frac{1}{2\pi b h}. \tag{D.19}$$

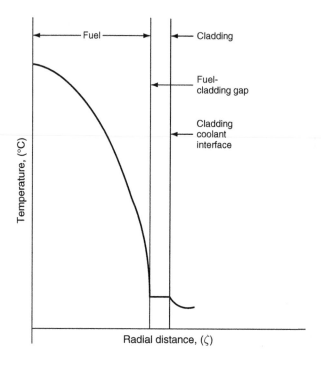

FIGURE D.2   Radial temperature profile for an oxide fuel element.

Likewise, replacing Eq. (D.6) with (D.7) we obtain the centerline thermal resistance:

$$R'_{cl} = \frac{1}{4\pi k_f} + \frac{1}{2\pi a h_g} + \frac{\tau}{2\pi a k_c} + \frac{1}{2\pi b h},$$

which satisfies

$$T_{cl} - T_c = R'_{cl}q'.$$

We can construct the temperature profile through the fuel element from the foregoing equations. The profile Fig. D.2 is typical for a ceramic fuel, such as the most commonly used UO_2, and a liquid coolant. By far the largest temperature drop occurs in the fuel, since its thermal conductivity is quite small. The thermal resistances thus are only weakly dependent on the fuel radius, and to a first approximation we may take $R'_{fe} \approx 1/8\pi k_f$. Under such circumstances the two thermal resistances are roughly related by $R'_{cl} \approx 2R'_{fe}$. Likewise, because the volume of cladding is small compared to that of the fuel, the mass used in determining the thermal time constant can to a first approximation be taken as that of the fuel.

APPENDIX E

Nuclear Data

TABLE E.1
Conversion Factors

1 eV	$1.6021892 \times 10^{-19}$ J
1 MeV	10^6 eV
1 amu	$1.6605655 \times 10^{-27}$ kg
	931.5016 MeV
1 W	1 J/s
1 day	86,400 s
1 mean year	365.25 days
	8766 h
	3.156×10^7 s
1 Ci	3.7000×10^{10} disintegrations/s

TABLE E.2
Physical Constants

Avogadro's number, N_A	6.022045×10^{23} mol^{-1}
Boltzmann's constant, k	1.380662×10^{-23} J/K
	0.861735×10^{-4} eV/K
Electron rest mass, m_e	9.109534×10^{-31} kg
	0.5110034 MeV
Elementary charge, e	$1.6021892 \times 10^{-19}$ C
Gas constant, R	8.31441 J mol^{-1}/K
Neutron rest mass, m_n	$1.6749544 \times 10^{-27}$ kg
	939.5731 MeV
Planck's constant, h	6.626176×10^{-34} J/Hz
Proton rest mass, m_p	$1.6726485 \times 10^{-27}$ kg
	938.2796 MeV
Speed of light, c	2.99792458×10^8 m/s

TABLE E.3
Microscopic Thermal Cross Sections (for selected isotopes, see also Table 3.2)

Atomic Number	Atom or Molecule	Symbol	Atomic Weight	Density (gm/cm³)	σ_a (barns)	σ_s (barns)
1	Hydrogen	H	1.008	gas	0.2948	47.463
	Water	H₂O	18.016	1.0	0.5896	99.52
	Heavy water	D₂O	20.030	1.1	0.0013	14.765
2	Helium	He	4.003	gas	–	0.79
3	Lithium	Li	6.940	0.534	0.0448	0.95
4	Beryllium	Be	9.013	1.85	0.0085	6.151
5	Boron	B	10.82	2.45	767	4.27
6	Carbon	C	12.011	1.6	0.0031	4.74
7	Nitrogen	N	14.008	gas	0.074	10.03
8	Oxygen	O	16.00	gas	0.0002	3.761
9	Fluorine	F	19.00	gas	0.0095	3.641
10	Neon	Ne	20.183	gas	0.039	2.415
11	Sodium	Na	22.991	0.971	0.354	3.038
12	Magnesium	Mg	24.32	1.74	0.0590	3.4140
13	Aluminum	Al	26.98	2.699	0.205	1.4134
14	Silicon	Si	28.09	2.42	0.152	2.0437
15	Phosphorus	P	30.975	1.82	0.146	3.134
16	Sulfur	S	32.066	2.07	0.518	.9787
17	Chlorine	Cl	35.457	gas	33.1	15.8
18	Argon	Ar	39.944	gas	0.675	0.656
19	Potassium	K	39.100	0.87	2.1	2.04
20	Calcium	Ca	40.08	1.55	0.38	2.93
21	Scandium	Sc	44.96	2.5	24.1	22.4
22	Titanium	Ti	47.90	4.5	5.68	4.09
23	Vanadium	V	50.95	5.96	4.47	4.95
24	Chromium	Cr	52.01	7.1	2.81	3.38
25	Manganese	Mn	54.94	7.2	11.83	2.06
26	Iron	Fe	55.85	7.86	2.56	11.35
27	Cobalt	Co	58.94	8.9	32.95	6.00
28	Nickel	Ni	58.71	8.90	4.49	17.8
29	Copper	Cu	63.54	8.94	3.35	7.78
30	Zinc	Zn	65.38	7.14	0.98	4.08
31	Gallium	Ga	69.72	5.91	2.56	6.5
32	Germanium	Ge	72.60	5.36	2.20	8.37
33	Arsenic	As	74.91	5.73	3.62	5.43
34	Selenium	Se	78.96	4.8	10.5	8.56
35	Bromine	Br	79.916	3.12	6.1	6.1
36	Krypton	Kr	83.80	gas	25.1	7.50
37	Rubidium	Rb	85.48	1.53	0.38	6.4
38	Strontium	Sr	87.63	2.54	1.28	10
39	Yttrium	Yt	88.92	5.51	1.13	7.66
40	Zirconium	Zr	91.22	6.4	0.185	6.40
41	Niobium	Nb	92.91	8.4	1.02	6.37
42	Molybdenum	Mo	95.95	10.2	2.52	5.59

Table E.3
(continued)

Atomic Number	Atom or Molecule	Symbol	Atomic Weight	Density (gm/cm^3)	σ_a (barns)	σ_s (barns)
43	Technetium	Tc	99.0	–	20.2	5.79
44	Ruthenium	Ru	101.1	12.2	2.56	6.5
45	Rhodium	Rh	102.91	12.5	9.39	–
46	Palladium	Pd	106.4	12.16	6.9	4.2
47	Silver	Ag	107.88	10.5	56.1	5.08
48	Cadmium	Cd	112.41	8.65	2233	5.6
49	Indium	In	114.82	7.28	193.8	2.45
50	Tin	Sn	118.70	6.5	0.603	4.909
51	Antimony	Sb	121.76	6.69	4.96	3.88
52	Tellurium	Te	127.61	6.24	4.6	3.74
53	Iodine	I	126.91	4.93	5.45	–
54	Xenon	Xe	131.30	gas	24.2	–
55	Cesium	Cs	132.91	1.873	2.3	–
56	Barium	Ba	137.36	3.5	1.2	3.42
57	Lanthanum	La	138.92	6.19	8.01	10.08
58	Cerium	Ce	140.13	6.78	0.58	2.96
59	Praseodymium	Pr	140.92	6.78	3.5	2.71
60	Neodymium	Nd	144.27	6.95	50.1	16.5
61	Promethium	Pm	147.0	–	–	–
62	Samarium	Sm	150.35	7.7	5025	38
63	Europium	Eu	152.0	5.22	4046	–
64	Gadolinium	Gd	157.26	7.95	43326	172
65	Terbium	Tb	158.93	8.33	20.7	6.92
66	Dysprosium	Dy	162.51	8.56	836	105.9
67	Holmium	Ho	164.94	8.76	57.3	8.65
68	Erbium	Er	167.27	9.16	138.6	9.0
69	Thulium	Tm	168.94	9.35	93.2	6.37
70	Ytterbium	Yb	173.04	7.01	30.8	23.4
71	Lutetium	Lu	172.99	9.74	66.4	6.70
72	Hafnium	Hf	178.5	13.3	92.2	10.3
73	Tantalum	Ta	180.95	16.6	18.2	6.12
74	Tungsten	W	183.66	19.3	16.3	4.77
75	Rhenium	Re	186.22	20.53	79.5	11.3
76	Osmium	Os	190.2	22.48	15.2	15.0
77	Iridium	Ir	192.2	22.42	376.8	14.2
78	Platinum	Pt	195.09	21.37	9.13	12.4
79	Gold	Au	197.0	19.32	87.42	7.90
80	Mercury	Hg	200.61	13.55	329.9	26.5
81	Titanium	Ti	204.39	11.85	3.04	10.01
82	Lead	Pb	207.21	11.35	0.137	11.261
83	Bismuth	Bi	209.0	9.747	0.030	9.311
84	Polonium	Po	210.0	9.24	–	–
85	Astatine	At	211.0	–	–	–

(continued)

Table E.3
(*continued*)

Atomic Number	Atom or Molecule	Symbol	Atomic Weight	Density (gm/cm^3)	σ_a (barns)	σ_s (barns)
86	Radon	Rn	222.0	gas	–	–
87	Francium	Fr	223.0	–	–	–
88	Radium	Ra	226.05	5	11.3	10.7
89	Actinium	Ac	227.0	–	–	–
90	Thorium	Th	232.05	11.3	6.51	13.5
91	Protactinium	Pa	231.0	15.4	177.8	33.0
92	Uranium	U	238.07	18.9	6.623	9.15
	Uranium dioxide	UO$_2$	270.07	10.0	6.623	16.8

Index